Constructing Community

Constructing Community

The Archaeology of Early Villages in Central New Mexico

Alison E. Rautman

The University of Arizona Press
Tucson

The University of Arizona Press
www.uapress.arizona.edu

Printed in the United States of America
19 18 17 16 15 14 6 5 4 3 2 1

Jacket designed by Leigh McDonald
Jacket photo: *Salinas Pueblo* by Birdie Jaworski

Unless otherwise noted, all figures were drafted by the author.

Publication of this book is made possible in part by a Humanities and Arts Research Program (HARP) Production Grant from Michigan State University.

Library of Congress Cataloging-in-Publication Data
Rautman, Alison E.
Constructing community : the archaeology of early villages in central New Mexico / Alison E. Rautman.
pages cm
Includes bibliographical references and index.
ISBN 978-0-8165-3069-4 (cloth : alkaline paper)
1. Pueblo Indians—New Mexico—Salinas Pueblo Missions National Monument Region—Antiquities. 2. Pueblo Indians—Dwellings—New Mexico—Salinas Pueblo Missions National Monument Region—History. 3. Pueblo Indians—New Mexico—Salinas Pueblo Missions National Monument Region—Social life and customs. 4. Farmers—New Mexico—Salinas Pueblo Missions National Monument Region—History. 5. Excavations (Archaeology)—New Mexico—Salinas Pueblo Missions National Monument Region. 6. Villages—New Mexico—Salinas Pueblo Missions National Monument Region—History. 7. Community life—New Mexico—Salinas Pueblo Missions National Monument Region—History. 8. Architecture, Domestic—New Mexico—Salinas Pueblo Missions National Monument Region—History. 9. Social archaeology—New Mexico—Salinas Pueblo Missions National Monument Region. 10. Salinas Pueblo Missions National Monument Region (N.M.)—Antiquities. I. Title.
E99.P9R18 2014
978.9'01—dc23
2014003134

♾ This paper meets the requirements of ANSI/NISO Z39.48-1992 (Permanence of Paper).

In memory of
my parents, Arthur and Emily Rautman,
and my friends, Jack and Louise Kite.
Thank you.

Contents

Illustrations

Figures

Unless otherwise noted, all figures were drafted by the author.

Tables

Preface

. . . the true ends of archaeology . . . are to recover and to piece together such mute evidences as time has spared of the careers of peoples who have left no written record of their achievements and of their failures; and to interpret those evidences in such a way as to teach us how and why man has become what he is today.

—Alfred V. Kidder, Introduction from *Prehistoric Southwesterners from Basketmaker to Pueblo*

In this volume, I review a lifetime of archaeological research in one geographic area, the Salinas Province of central New Mexico. Over the last twenty-something years I have directed several excavations at four different archaeological sites, ranging in time from about A.D. 900 until the mid-1300s. I have also participated in excavations at some of the later pueblos that were part of the Spanish mission system in this region and have assisted colleagues with other projects. The research reported here synthesizes information from some of my published works, together with information from unpublished papers that I presented at professional meetings and from limited-distribution descriptive site reports. I include much unpublished data from long ago, some reanalyses of old data, and also some results from my recent excavations.

I place a number of different specific projects into a larger research framework that encompasses a very broad topic: the development and organization of early village communities in the American Southwest. In this volume, I discuss explicitly the logic, linkages, and lines of thought that weave together many specialized analyses and approaches in order to reconstruct, as Kidder says above, what life was like for ancestral Pueblo Indian people in one geographic area, to trace how their lives changed over the centuries, and to examine how their experiences and history form part of our collective humanity today. The basic question here asks about the connection between what we know about the social life of people who lived in towns, villages, and other small communities, and what we see in the archaeological record.

Acknowledgments

The research reported in this volume was supported by a number of different grants, a large number of student fieldworkers, and the sustained help and support of innumerable individuals. Most importantly, of course, my parents, Arthur and Emily Rautman, encouraged and supported my interests in archaeology. They truly loved everything about New Mexico's landscape, history, and culture, and they greatly enjoyed taking part, even vicariously, in my archaeological adventures.

I also owe a tremendous debt to Jack and Louise Kite, whose ranch included the historic town of Gran Quivira as well as two of the archaeological sites discussed in this volume. Much of the research reported here was made possible by their interest in local archaeology, their respect for site conservation and preservation, concern for community issues, and their support of education and research. They were remarkable people, and I am glad to have known them.

My thoughts about early village organization in Salinas have been shaped by innumerable discussions and adventures with Kate Spielmann, and also with Billy Graves, Tiffany Clark, Matt Chamberlin, and Julie Solometo. This particular manuscript was also improved by specific discussions with Matt Chamberlin and also Tom Rocek, both of whom generously read and corrected mistakes in selected chapters throughout various iterations of the manuscript. However, none of these colleagues can be held responsible for any errors I have made in reading, reporting, or interpreting their research.

Research at Kite Pithouse Village was made possible by funding from several sources, including a Dissertation Improvement Grant from The University of Michigan Rackham Graduate School, a grant from The University of Michigan Department of Geological Sciences Turner Fund, a grant from The University of Michigan Museum of Anthropology Griffin Fund, and a dissertation improvement grant from the National Science Foundation (with my advisor John Speth). Funding from The Wenner-Gren Foundation for Anthropological Research and from Sigma Xi, The Scientific Research Society, helped pay for the specialized analysis. The writing of the dissertation itself was assisted by a Horace H. Rackham Predoctoral Fellowship and by a predoctoral fellowship from the American Association of University Women.

My volunteer crew at Kite Pithouse Village included graduate and undergraduate students from the University of Michigan: Jody Brown, Ross Harper, Holly Houghton, Karen Leidel, Lindy Mendelsohn, Margaret Sattersmoen, Rasmi Schoocongdej, Tristine Smart, J. Homer Thiel, and Jennifer Weisberg. Tom Carroll, then superintendent of Salinas Pueblo Missions National Monument, generously provided room for a field lab at the National Park Service facilities in Mountainair, and helped us find housing in town. I am also grateful to my advisor John Speth for a professional lifetime of assistance; he and Tom Rocek helped with coring the site. I thank Karen Leidel particularly for adopting our field cat, Claunch.

The archaeological surveys reported in the chapter on jacal sites were not my projects; I thank Kate Spielmann and Matt Chamberlin for the opportunity to accompany them on their survey projects around Gran Quivira. All of the information presented about their survey research is taken from their various publications (cited in the text), from my own personal observations, and from extensive discussions with them, but neither of them can be held responsible for any errors of fact, faulty research, or for any interpretations that I present here as my own.

Excavations at Kite Pueblo in 1994 were funded by grants from The National Geographic Society and from The Wenner-Gren Foundation for Anthropological Research. A postdoctoral fellowship from the American Association of University Women was critical in supporting my analysis and writing. As usual, the staff at Salinas Pueblo Missions National Monument provided practical as well as moral support, and helped us find housing at the Gran Quivira unit.

The field crew at Kite Pueblo involved a heroic group of volunteers: David Bruining, Chris Carbone, Aimee Centevanny, Chris Corner, Mike Havens, Nicole Kowrach, Amanda Ledwon, Charles Mathur, Dan Meyer, Dominic Perrone, Emily Pudden, Mary Pyott, Carlie Rodriguez, Eva Ros, Susie Rynkiewicz, and Nancy Snyder. Teresa Kugler was particularly valuable as my unpaid field director. Wayne Bischoff helped with much of the analysis as my research assistant during the 1994–1995 academic year.

Excavation at Pueblo de la Mesa in Cibola National Forest during 1992 and 1993 was funded by grants from the US Forest Service as well as from The National Geographic Society. I appreciate Joseph Tainter's assistance in shepherding my project through the US Forest Service permit process.

Because of the history of illegal digging at the site, I received authorization to remove the burials as they were encountered.

Excavations at Pueblo de la Mesa involved students who were enrolled in the Michigan State University archaeological field school: Barbara Alldaffer, Pamela Bird, Jennifer Chapp, Ellen McAndrew, James Montney, Eileen Moran, Jennifer Nisengard, Mary Pyott, Scott Rembrandt, and Jill Riolo. Glenn Stuart and Teresa Kugler helped supervise; volunteers Diane Rice and John Lawrence also helped with excavation and innumerable other tasks.

The second-year field school included Paul Blonsky, Aimee Centevanny, Josh Fredrickson, Amanda Ledwon, Rebecca Margolis, Anna Naruta, Dominic Perrone, Emily Pudden, Todd Reck, and Sarah Wilkins, supervised by Amy Hirshmann, Anne Colyer, and Teresa Kugler. Diane Rice, Sharon Hanna, and John Lawrence provided invaluable volunteer labor. The US Forest Service Passports in Time program also provided a crew of volunteers for two weeks, supervised by Cynthia Benedict and Scott Baker. I thank Delbert and Helen Towell for letting us use the ancestral Devaney homestead for our field headquarters, and for welcoming us into their lives. They went far beyond the call of duty in assisting with construction, cooking, baby care, car repair, and innumerable other tasks. I could not have managed these two field seasons without their practical and moral support.

Research at Frank's Pueblo is based upon work supported in part by the National Science Foundation (Grant #0521928). Additional funding and labor came from students enrolled at the 2007 and 2008 archaeological field schools through Michigan State University and through James Madison University (Virginia). Matt Chamberlin of JMU provided invaluable help in the field as my codirector. Student and volunteer crew members in 2007 included Natasha Arnold, Caitlin Bauer, Chantee Brakeville, Katie Chinn, Robert Cinninger, Samantha Fladd, Courtney Fountain, Meghan Jacokes, David Kolkema, David Lewandowski, Aaron Moss, Beth Pruitt, Chris Samoray, Grayson Shephard, Caryn Tegtmeyer, James Thorpe, Erica Dzrudzk, Marjie Heyman, and Diane Rice. Marieka Brouwer Berg, Nicole Burritt, Jacob Foss, Chris Stawski, and Chris Valvano helped supervise in the field; Jake Foss was my research assistant during the next academic year. His detailed analysis of the lithic sample (Foss 2009) became part of his master's thesis from MSU.

The 2008 crew included Travis Altomonte, Elizabeth Bikowski, Kiel Chamberlain, Zachary Copley, Jeff Gepper, Shaunah Gordon, Grant

Hayes, Clare Heisey, Marjie Heyman, Jocelyn Mallin, Harlan McCaffery, and Daniel Perkins. Marieka Brouwer Berg, Josh Moss, and Katie Hammond Frederick helped supervise the students. As usual, Diane Rice provided volunteer labor and a lifetime of friendship. Marieka Brouwer Berg was my research assistant during the 2008–2009 academic year; among other tasks, she took the lead in compiling and editing the field report for the 2008 field season. As part of a comparative study of all the Salinas sites, Marieka helped compile (finally) a comprehensive descriptive report from my dissertation excavations at Kite Pithouse Village (Brouwer and Rautman 2013), and Marjie Heyman reanalysed and compiled much unpublished data from Kite Pueblo (Rautman and Heyman 2011).

I also thank all the people who helped with the details of organizing and operating different field projects, including the many residents, public officials, and business owners in and around Mountainair. I am particularly grateful for all of the practical help, housing, and moral support that Glenn Fulfer and other people at Salinas Pueblo Missions National Monument have provided over the years. I also greatly appreciate their willingness to allow us to consider the Park Service as our emergency backup system—no small thing in the days before cell phones. My brother Chris Rautman moved his family to Albuquerque soon after I began working in New Mexico; little did he suspect he had also signed up for providing so much volunteer labor, housing, and mileage in support of my research. Susan Brazil and Tracy Kite helped enormously during the 2007 field season, when we lived in the historic town of Gran Quivira on the Kite ranch. I particularly cherish the time I had to visit with their grandmother, Louise Turner Kite, during her last summer.

In broader terms, I have always appreciated the faith placed in me by my mentors at the University of Michigan, John Speth and William Farrand. Michael Jochim (University of California–Santa Barbara) generously overlooked my lack of collegiate field school experience, allowing me to participate in his excavation in Germany and to write my master's thesis on the archaeological geology of the site.

I was trained in Southwestern archaeology at the Washington State University archaeological field school, affiliated with the Dolores Project at what is now Crow Canyon Archaeological Center in Dolores, Colorado. I thank Bill Lipe and Tim Kohler of Washington State University for taking a chance on me, a latecomer to archaeological fieldwork. Kim Smiley similarly gave me the benefit of the doubt (and a job) with the Black Mesa Archaeological Project in Arizona.

Kate Spielmann (now at Arizona State University) introduced me to Salinas archaeology during two summers of fieldwork at Gran Quivira Pueblo with her University of Iowa archaeological field school; I have been very fortunate indeed to have such a good mentor and lifelong friend. Jane Kelley (University of Calgary) and Regge Wiseman (Museum of New Mexico) each very generously spent several days with me after that field school, providing guided tours of various pithouse sites in the Sierra Blanca and Salinas areas, and discussing numerous potential dissertation topics.

My geology advisor Eiler Hendrickson at Carleton College first encouraged my efforts to combine concepts from geology and questions from archaeology; Tim Vick made it possible for me to do so by helping me with the technical aspects of my B.A. thesis regarding the mineral and chemical composition of European Bronze Age slags. I owe them both much more than the six-packs of beer I gave them in token repayment.

Going *much* further back, however, I also owe a great debt to the Girl Scouts of America (GSA). In particular, the GSA sponsored the Wider Opportunity adventure "Dig Mankind" at National Center West (NCW) in Ten Sleep, Wyoming, which I attended in 1974. Caroline Adams, a graduate of the University of Wyoming, directed the project and inspired me by her example. I then worked at NCW every summer through college, teaching backpacking and archaeology. Although early in my career I sometimes regretted my lack of formal field school experience, in fact, the skills I learned in Girl Scouting more than prepared me for nearly everything I needed to know about archaeological fieldwork that was not covered in graduate school: how to build a latrine, fish valuables out of a latrine, change a tire, stop the bleeding, deal with crises, and organize volunteers. I will be forever grateful.

Finally, I am particularly indebted to my husband John, who patiently put up with it all and generally kept the home fires burning. Rudy and Ingrid probably sacrificed the most to make this research possible; I thank them for lending their mom to Archaeology for so many field seasons.

Constructing Community

Introduction

Interpreting Archaeological Village Sites

In this volume, I examine some of the issues involved in linking the inanimate world of archaeological remains with the living world of people and their everyday activities. The evidence that we see in the archaeological record represents the ideas and behavior of individual human beings, of course, but not in any direct sense. The challenge of archaeological research involves just this problem of interpreting the evidence that we have in light of our ideas about individual people, social groups, human beings in general, and also about the past. Here, I examine particularly the concept of the village, a word that has both physical and social connotations. While many archaeologists, including myself, may casually interpret a given site as a "village," there is no particular reason why we should do so. In fact, how do we know if a village site really does represent a village?

The research described here investigates this relationship between an archaeological site that appears to be a village with the anthropological concept of the village as a social group, a community. In my case study from the American Southwest, this research begins with observed evidence from the built environment. Southwestern archaeologists have long argued that specific architectural features such as plazas or kivas act as "integrative structures," and that they have a role in fostering and expressing social connectedness among different groups of people (e.g., Lipe and Hegmon 1989; see also discussion in Lekson 1998, 2008). In contrast, my focus on the built environment is broader: it includes all of the structures, features, pathways, and open spaces in a particular locale on the landscape, as well as the spatial relationships among them.

We know that the built environment as a whole defines people's use of space in a way that can both express and also limit social interaction (Doxiadis 1968; Rapoport 1969; Hillier and Hanson 1984; Lefebvre 1991). Buildings and other architectural features can demonstrate social and political power (Weeks 1988; Nielsen 1995; Hegmon et al. 2000; Reese-Taylor and Koontz 2001; Russo 2004; Inomata 2006; Lahiji 2011). The built environment in general can also facilitate ritual events, express ritual concepts, and communicate worlds of meaning (Rapoport 1982, 1988; Ashmore 1991; Shafer 1995; Psarra 2009). Specific architectural features as well as the entirety of the built environment can also link the secular world to the sacred by situating the community within sacred cosmological space (Moore 1996; Van Dyke 2008, Fowles 2009; Nelson et al. 2011).

The built environment can also separate individuals and spatially define and reinforce distinctions between social groups. Plazas and passageways can organize structures into defined spatial clusters, which can be associated with, or crosscut, previously established social dimensions. Social status can be indicated by a room's placement in relationship to exterior spaces such as plazas or in relationship to other buildings (Low 1995, 2000; Inomata 2006). Encounters between individuals or groups may also be encouraged, enforced, limited, or even prohibited by the size, shape, and location of exterior as well as interior spaces (Moore 1996; Shapiro 2005). In all these ways, the built environment helps create, express, and also reinforce a shared "public" or official culture (Lawrence and Low 1990; Low 2000; Creel and Anyon 2003).

Of course, not all important cultural concepts, interactions, or social groupings will find expression in the built environment. However, it is also true that evidence of the size, shape, spacing, spatial organization, and arrangement of the built environment does tell us something about the everyday activities, patterned interactions, and enduring social categories experienced by groups of people who lived in that locale, and also about the individuals who actually sat, walked, worked, and thought in these spaces (McGuire and Schiffer 1983; Zeisel 1984; Brand 1995). These individuals are the ones who constructed the buildings, remodeled them, and used them, but they did so within the context of a social group that shared many—but not necessarily all—of the same ideas, assumptions, expectations, and values over some period of time (e.g., Rapoport 1969). In this study I investigate how people, individually and together, constructed their village as an archaeological site and also as a social community.

Village Sites and Village Society

The appearance of the earliest village sites in the archaeological record signals a very widespread economic, social, and demographic transition. The earliest village sites developed during the Near Eastern Neolithic period about eight thousand to ten thousand years ago, but the dates of similar types of sites vary in different regions across Africa and Eurasia (e.g., Kohler et al. 2008; Bocquet-Appel 2009; see also Kuijt 2002; Robb 2007). In the Americas, the comparable period of early village settlements dates much later in time, ranging between about 1500 B.C. to A.D. 500. In the Americas, this time of early villages was originally called the Formative period (Willey and Phillips 1958). This terminology is still common in Latin American archaeology, where, for example, the Formative period precedes the Pre-Classic, Classic, and Post-Classic periods, which are broadly defined and distinctive periods of time that are distinguished by very large, politically unified geographic regions and by the development of conquest states and empires.

This kind of progressive framework may give the impression that early villages are in the process of development, or that they are the source or origin of something else. And indeed, some archaeologists do look to these small-scale societies to investigate the earliest glimpses of incipient political complexity and the emergence of political elites (see Lightfoot and Upham 1989; Nelson 1995; Lekson 2006a, 2006b). However, this perspective does not do justice to the variety of ways that people organized and interacted with one another in small villages, and how they coped with a myriad of social, political, and economic challenges in many different areas around the world.

In any geographic region, the periods characterized by early villages are generally recognized as a time during human antiquity in which relatively small groups of people began staying for an extended time in one locale. Early villages of the Neolithic or Formative eras, as a group, differ from later towns, villages, or hamlets of similar size that existed within feudal, peasant, or colonial systems of governance. Archaeologists who study early villages thus frame their discussion in terms of comparisons and contrasts with the mobility strategies, economic choices, and social dynamics that we associate with small-scale, basically egalitarian forager groups (e.g., Kohler et al. 2008; Bocquet-Appel 2009).

Archaeologists generally recognize the formation of early villages as one aspect of an overall decrease in residential mobility among foragers

(Binford 1980, 1990). In this sense, clusters of domestic structures formed the physical expression of people's social, political, and economic activities concentrated in a particular locale. Even a brief glance at the vast ethnographic literature on forager societies (e.g., Kelly 1992, 1995; Barnard 2004) suggests that such a change in individual and group mobility is likely to have significant social, political, and economic impacts, as well as impacts upon the local and regional landscape (Beck et al. 2007; see also Price et al. 2000).

Village sites themselves usually include a number of residential structures or features. Areas of waste disposal and of disturbed sediment characterize repeated or lengthy human occupations, affecting the local plant and animal ecology. More elaborate or permanent structures and features impose a spatial organization within the built environment, channeling foot traffic as well as affecting people's experience of the local landscape through sight, sound, and smell. In particular, given what we know of forager lifeways, we can expect that any significant change in mobility and land use will likely also challenge established strategies and cultural rules of interpersonal conduct, conflict management, and social relationships both within and between local groups.

However, there is considerable diversity within archaeologically observed village sites and many different ways of "living in a village." Some residential sites may represent repeated seasonal or short-term camps; others may record the activities of a larger or more permanent population. Some sites show that people cultivated domesticated plants, but again there is considerable variation in how much people depended on farming. Thus, while we recognize early village sites across the globe, it is also reasonable to expect to see considerable variation among them in overall group mobility, economy, and social dynamics.

Despite these observed variations, early village sites in the archaeological record seem similar enough to justify examining them as a more general worldwide phenomenon (e.g., Bandy and Fox 2010). The subject of early village society as a topic for archaeological research is not a new one (e.g., Willey and Phillips 1958). Recent studies, including this one, differ by placing greater emphasis on the social processes involved in this relatively long-term change toward village formation in many parts of the world, and the diversity of social strategies represented among these cases (e.g., Canuto and Yeager 2000; Bandy and Fox 2010). In these particular studies, the "village" is not necessarily seen as a natural or inevitable social unit, but as a contingent array of interacting individuals, groups, and factions (Isbell 2000). Following the usage outlined by these more recent scholars, I consider that early village society represents a dynamic concept rather than a distinct social type or stage in cultural evolution.

Village societies in general thus represent a widespread and long-lived form of social and political organization, a human invention that provided a flexible, versatile, and resilient plan for coping with long-term group interactions in a context where some of the tried-and-true social strategies, developed among foraging groups, were no longer possible. While some scholars do find village society interesting simply as a starting point for the development of social and political inequality, most contemporary scholars take a more nuanced view of social concepts such as "complexity." The study of early village society thus includes considering the social complexities inherent within basically egalitarian society. These complexities include a wide range of possible organizational frameworks and group dynamics through which people managed to develop the various "institutions, practices, habits of thought, action, and expression" that together comprise village life (Bandy and Fox 2010:16).

Interpreting Archaeological Village Sites

Although many archaeologists, including myself, casually talk about "village sites," interpreting the material remains of a possible village should really involve a much more complicated process of reasoning. Archaeologists may map and describe a village site based on surface indications of a couple of structures or features and some artifacts, but these remains may actually represent a series of short-term campsites, utilized by one or two families at a time—not really what one thinks of as a village.

The basic research questions of this study concern just this issue: Given an archaeological site that seems to have some residential structures, how can we tell if this site really represents a village? And if we decide that the archaeological site should be considered a village, then how did that village function? How were the people organized? To what extent did people who lived there see themselves as a group, and how did individuals and smaller social groups interact within the village as a whole? These questions ask us to articulate the connection between evidence that we see in the archaeological record and the social processes and dynamics involved in a village as a living community of interacting individuals.

As a practical matter, in order to keep these distinctions clear, I will apply the concept of "village society" to the general social organizational dynamics described in recent archaeological studies, such as those included in Bandy and Fox's (2010) edited volume. The concept of "early village society" concerns the more comprehensive and comparative study of the similarities and differences among case studies from Neolithic or Formative time periods: village societies within a world of villages, as opposed to small settlements that were associated with much larger political organizations such as empires or colonies (see also Lightfoot and Martinez 1995; Parker 2006).

I will reserve the term "village site" to refer to a specific locale where a field researcher has identified one or more possible residential structures. In the American Southwest, these structures might include pithouses and/or surface rooms with walls made of wattle and daub, adobe, or masonry. As we develop an idea or mental model of a site's size, spatial organization, and appearance based on evidence of structures, features, and also artifacts, we also interpret people's experience of life within a built environment as the local expression of individual acts and social dynamics within a specific local history and context (e.g., Ortman et al. 2000). This terminology emphasizes that a "village site" is located in a specific place, but is also an idea or mental model that we are using today to think about the visible record of the past, whether we are visitors, tourists, or professional archaeologists.

A second problem in studying early village society involves another kind of model building: that is, identifying and interpreting archaeological evidence of these specific and local practices. This process involves inferring the internal organization and dynamics of social interactions among those people living in the village site itself. As one might expect from such a widespread and long-lasting form of social organization, specific village societies are likely to exhibit considerable variation in their internal organization and the social relationships between them (Wills and Leonard 1994). Understanding this variation involves identifying what social groups were important, as well as the size and composition of these groups. Were all these groups equally participating in social interactions at the village site? Did people live in nuclear or extended families or lineages, and how can we tell (Dozier 1965; Haviland 1988)? What was the range of variation in family size and composition over time (Goody 1971), and how did these groups relate to our conception of households (Wilk and Rathje 1982; Wilk and Netting 1984)? Were other kinds of corporate groups more important? How were these groups defined, and what did they actually do? Did all village members cooperate in certain tasks or were village subgroups more important in organizing labor? As one example, Bernardini

(2000) has found that "kiln firing groups" may have been of importance in structuring interactions between different households. Did subgroups such as these actually play much of a role in daily life, or was it mostly a case of "every household for itself"? In either case, how did people cope with the vicissitudes of life or troubles stemming from bad luck, poor health, poor harvests, or poor judgment?

All these questions concern the internal dynamics of each village. While we can think of early village societies as a global and long-lasting social invention, understanding the social interactions in a given village site involves an understanding of the local social group, the community, within its own context and history (e.g., Crown and Kohler 1994; Cameron and Duff 2008). I will reserve the term "village community" to convey the idea of these local social dynamics that we hope to infer from the archaeological evidence of a particular site that we think of as a village.

This conceptual step highlights an important caveat. Indeed, just as a collection of individuals does not necessarily represent a coherent group (e.g., Palmer et al. 1997), it is also true that an archaeologically observable set of domestic dwellings in a particular locale does not necessarily mean that this archaeological site represents a community. Instead, the organization and operation of this community is something that we must figure out, starting with the evidence that we have from a particular site itself and from what we know of the other various archaeological sites within a given region.

In this volume, I investigate the connection between archaeologically defined village sites and their organization as social communities in one area of the American Southwest—the Salinas region of central New Mexico. In doing so, I am interested particularly in integrating what we know of living communities and community organization with the archaeological expression of village life. If we do think that our archaeological site represents a village, then how did that village community work? What was early village society like for people at that place in the past? What aspects of life represent a broadly shared experience, or alternatively, which were unique to one particular time and place and cultural tradition? What does the archaeological record tell us about an individual's life experiences, and how did those experiences affect the organization and operation of different social groups within the village and also the organization of the village as a whole? How widely shared were these sorts of experiences and strategies? Were they specific to the people who lived in a particular village, or were they similar to those within a broader geographic region?

What, if anything, do these findings tell us about small-scale village society in general? The research reported here begins to address some of these questions.

Early Village Sites in the American Southwest

In the ancient American Southwest, the first villages of semisubterranean houses record a change in mobility that occurred over a span of time from about A.D. 200 to 900 (Cordell and McBrinn 2012), a change that is very loosely associated with other changes in the use of pottery and farming of domesticated plants. In the Pecos Classification System chronology (Kidder 1924), these earliest sites are sometimes called Basketmaker sites because at the time, archaeologists thought that they predated the development and use of pottery. The residential structures (pithouses) in Basketmaker sites housed people who we might describe as hunter-gatherers or mobile foragers/collectors. In some cases, they may also have relied on cultivated maize, and some even used pottery as well (Glassow 1972; Wills 1988a, 1988b; Matson 1991; Gilman and Young 2005). Although these pithouse dwellers were probably quite mobile, during at least some parts of the year, they were sedentary enough to justify building reasonably durable shelters (Cordell and McBrinn 2012).

Archaeologically observable pithouse village sites across the American Southwest generally show evidence of one or more residential structures and associated features, indicating that a small group of people occupied that particular locale (e.g., Whalen 1994a, 1994b). In some cases, thick midden accumulations of domestic trash and organic debris show that some locales were occupied for lengthy periods of time, or possibly repeatedly reoccupied by multiple family groups over several years or even generations. This kind of occupation shows considerable social continuity and organization, even if the particular locale was not continuously occupied. In other cases, it is not clear whether these sites are really villages at all, or if they represent the chance spatial clustering of unrelated short-term occupations. The continuing debate about the meaning and significance of Shabik'eschee Village, an early pithouse site in Chaco Canyon, illustrates the importance of this distinction between the archaeologically defined village site and our interpretations of local social dynamics (see discussion in Wills et al. 2012).

Later pithouse village sites in the Southwest show abundant evidence of pottery and also domesticated plants such as maize (Wills 1992; Hard et al. 1996; Doolittle and Mabry 2006). In many parts of the Southwest, pithouse village sites such as these were common by about A.D. 600 (Wilshusen 1991). Although some people continued to use pit structures for houses much later in time, in general by about A.D. 900, most residential sites are dominated by construction of aboveground houses or pueblos (Cordell and McBrinn 2012). Among Southwestern archaeologists, both kinds of residential sites—pithouse and the earlier pueblos—are considered early villages, occupied at least seasonally by groups of people with some sort of shared history, ideas, and expectations (e.g., Varien and Potter 2008; Kohler et al. 2010; Kohler and Varien 2012; Wilshusen et al. 2012).

It is important to note, however, that some pueblo sites became very large indeed, with many hundreds of contiguous and interconnected rooms forming highly compact and nucleated settlements. In some parts of the Southwest, large pueblos also cluster together in a given locale, recording complex interactions and relationships within even larger regional populations (e.g., Mera 1933, 1940). Ideas about early village society considered in this study may not be appropriate to this much larger scale of local and regional dynamics.

Indeed, some scholars have questioned whether the observed historic and ethnographic record of Pueblo Indian society in these large towns after Spanish conquest in the sixteenth century is a useful analogy for understanding early pueblo organization at all (e.g., James 1997; Lekson 2009). Others (e.g., Liebmann and Preucel 2007) point specifically to the impact that Spanish culture and institutions had on Pueblo Indian groups, and on ways that people coped with and actively resisted Spanish influence and domination (Spielmann et al. 2008). In other studies, concepts of political organization that are not necessarily associated with modern Pueblo Indian society have helped archaeologists understand the organization of regional social networks among the more ancient pueblos (e.g., Spielmann 1994b, 2004b).

These caveats are important here because several of these large pueblos that were occupied in the sixteenth and seventeenth centuries are also found in my study area. Three of them—Abó, Quaraí, and Gran Quivira Pueblos—are preserved today as Salinas Pueblo Missions National Monument. These three sites dominate the visible archaeological landscape with their large and extensive house mounds, as well as the impressive ruins of the Spanish mission churches. Because archaeological and historical studies of these large pueblos have been so significant to our understanding of Pueblo Indian life in central New Mexico, any study of early villages in this area must begin with an understanding of the most visible of these archaeological sites, their geographic setting, and their place in the local culture of the people who live near the archaeological sites today. It may be tempting to trace a direct connection from Spanish descriptions and evidence of Indian society and lifeways in the sixteenth and seventeenth centuries to practices in the more distant past, but it is also important to realize any such relationships should be established through archaeological research rather than simply assumed. It is also important to note that the social dynamics within and between these large late pueblos may not be directly relevant to my study of early village life in considerably more ancient times.

The Salinas Province of Central New Mexico

Back of the Sierra Morena [Monzanos] there were found some salines, which extend for five leagues, and which produce the best salt that Christian people have seen. It resembles nothing but salt derived from the sea. In them [the salines] five pueblos were discovered on the slopes of the Sierra Morena.

—Hernan Gallegos, from George P. Hammond and Agapito Rey, *Narratives of the Coronado Expedition*

My research area is located in the geographic center of New Mexico, an area called the Salinas Province (figure 1). This area lies between the traditional culture areas of the ancestral Rio Grande pueblos to the north and highland Jornada Mogollon groups in the Sierra Blanca Mountains to the south. The archaeological record of the Salinas Province includes evidence of cultural contact with groups from around the Zuni area in western New Mexico, as well as with the closer Rio Grande pueblos. The Indians of Salinas were also interacting with Plains Indian groups to the east; Spanish documents record how bison-hunting Indians from east of the Pecos River came into the Salinas region to trade bison meat and hides for maize and cotton (Hayes et al. 1981; Hayes 1993; Hickerson 1994).

The Salinas Province is probably one of the lesser-known parts of the Southwest among archaeologists today (Tainter and Levine 1987), especially compared to other regions such as Chaco Canyon or Mesa Verde in the northern Southwest (see figure 1). However, early explorers, traders, and also archaeologists such as Adolph Bandelier wrote extensively about the large Indian pueblos of the area (Gregg [1844]1967; Carleton 1854; Bandelier [1881]2008, 1890; Hammond and Rey 1940; and see

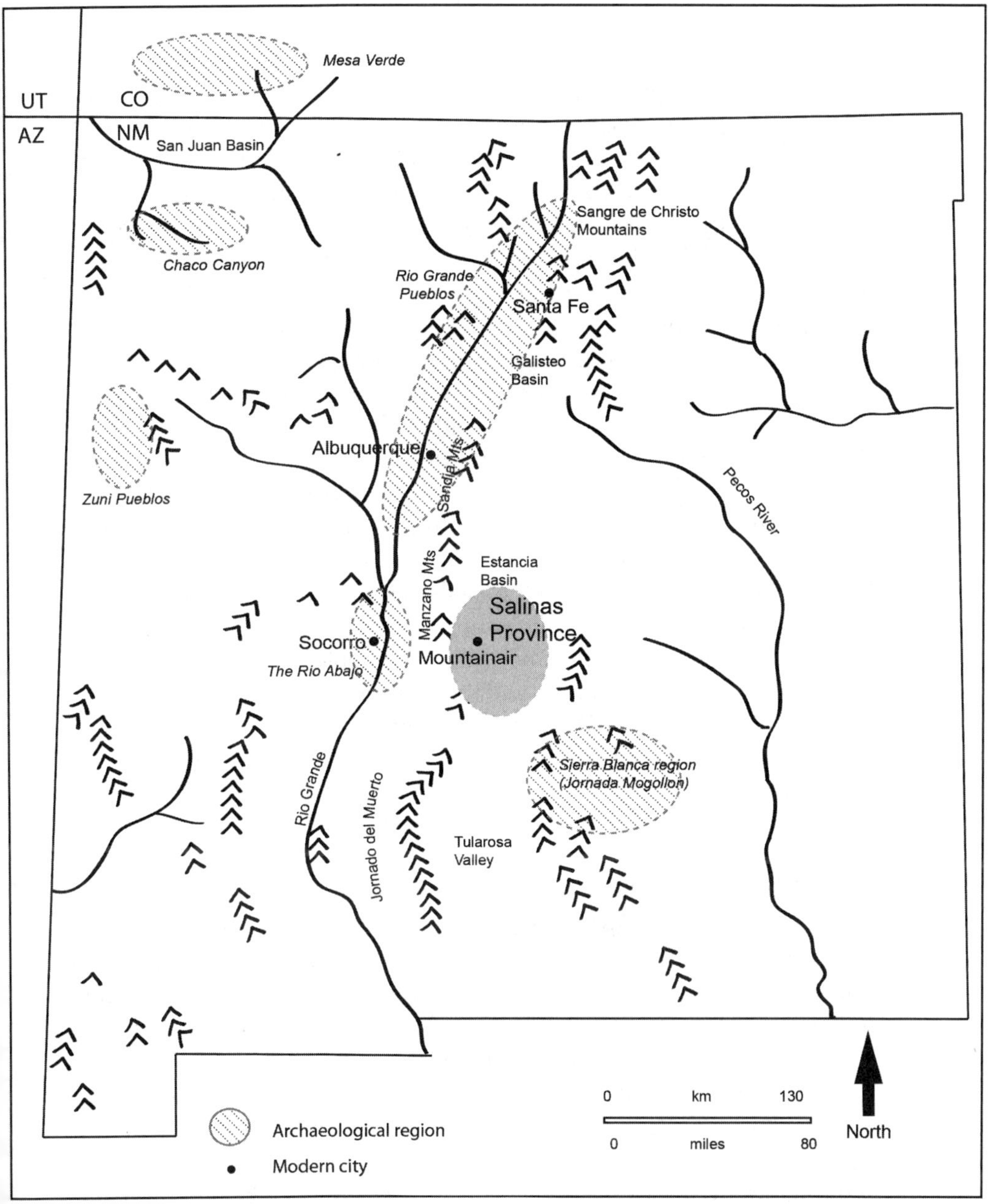

Figure 1. Location of the Salinas Province within New Mexico. Hatchured areas show the approximate locations of other significant Pueblo Indian regions considered in this volume, including the Rio Grande Pueblos (the Eastern Pueblos), the Zuni area, the area around Chaco Canyon and Mesa Verde, the Rio Abajo area near the modern town of Socorro, and the Sierra Blanca region of the Jornada Mogollon, located in upland areas to the south of the Salinas Province.

summaries in Forrest 1929; Spicer 1961; Lange and Riley 1966; Flint and Flint 2005). This area is also remarkably beautiful, with gleaming white salt lakes tucked within gently rolling sandy plains. Dusky red sandstone of the Abó formation forms nearly vertical slopes on the north rim of Chupadera Mesa by the town of Mountainair, and on the western edge of the Estancia Basin, the pine forests of the Manzano Mountains create a dark green backdrop for the entire region.

The name "Salinas" comes from the Spanish explorers and conquerors of the region and refers to the salt lakes (*salinas*) of the Estancia Basin. Probably because of the economic importance of the salt deposits to the Spanish empire, and the well-preserved evidence of the Spanish occupation, much of the archaeological research in the Salinas region has focused on the very large late pueblos where the Spanish Franciscan priests established mission churches. The archaeological sites of Salinas Pueblo Mission National Monument are administered from the National Park Service (NPS) headquarters in the modern town of Mountainair; they are shown in figure 2. Although discussion of these large late sites may seem somewhat tangential to the study of early pueblo villages, the archaeological record of these pueblos, and the Spanish descriptions of life during the colonial mission period are important in framing many archaeological expectations regarding the Pueblo Indians in Salinas.

Salt and Water

The salt deposits that loom so large in the ancient and more recent history of this region resulted from evaporation within the internal drainage of the Estancia Basin. The salt lakes began forming thousands of years ago at the end of the Ice Ages (Bachhuber 1982). During the Pleistocene, between about twenty-four thousand and twelve thousand years ago, a large pluvial lake that geologists call Lake Estancia filled the basin. This freshwater lake measured up to 46 meters deep and 1,165 square kilometers in size (Meinzer 1911); the modern towns of Estancia and Willard would have been covered with at least 30 meters of water. After the Ice Ages, about ten thousand years ago, a second and smaller pluvial lake, Lake Willard, formed in the same area. Minerals from the underlying rock, the Yeso formation, dissolved into the lake water, becoming more concentrated as it evaporated (White 1994; Allen and Anderson 2000; Allen 2005).

About 7,000 years ago, the regional climate became warmer and drier. Higher rates of evaporation together with a decrease in precipitation led to decreasing groundwater levels. Ephemeral playa lakes across the region

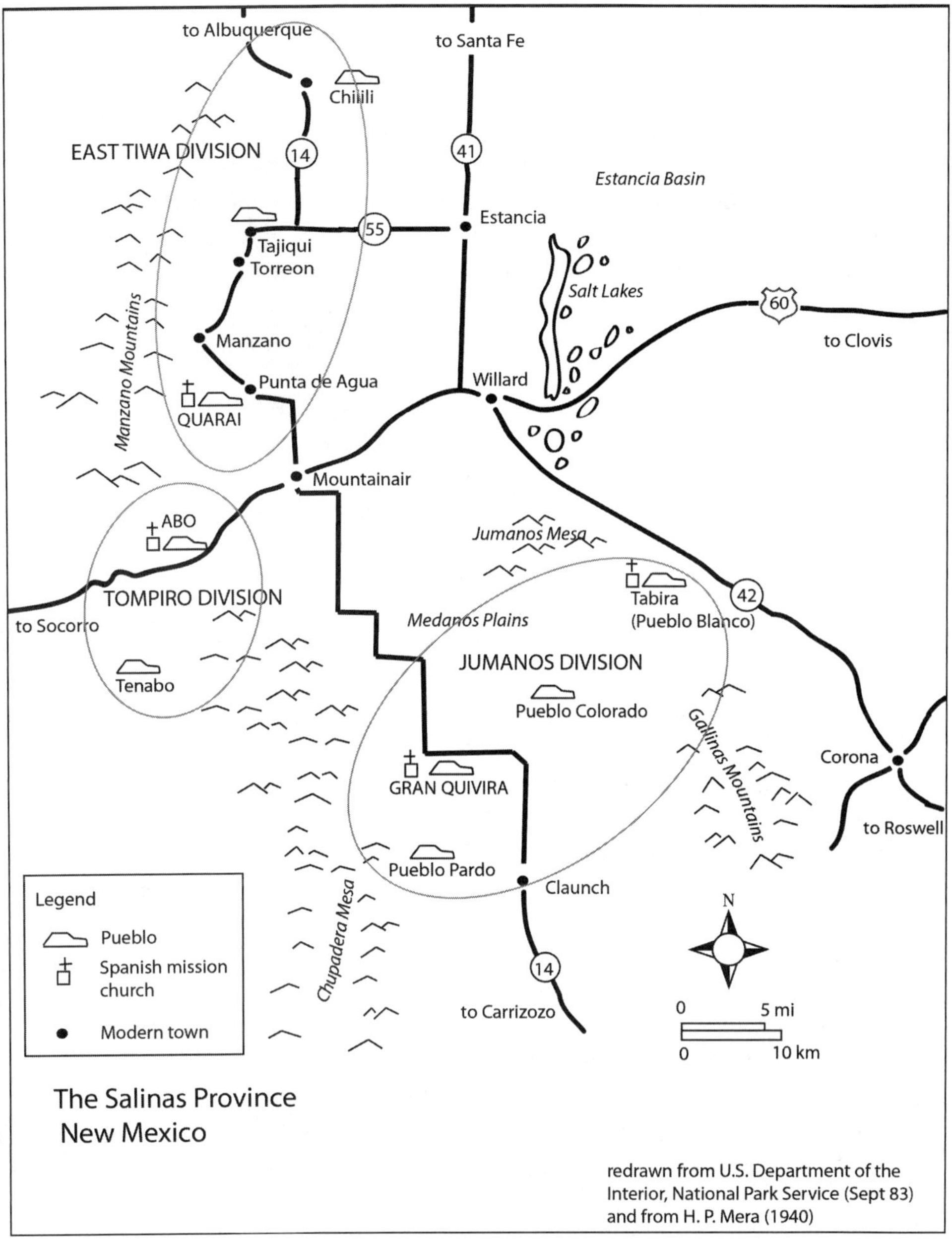

Figure 2. Prehispanic and historic period pueblos of the Salinas Province. The circled areas show the linguistic subdivisions of the Salinas Province used by H. P. Mera (1940) in his study of Rio Grande glaze ware ceramics. The research discussed in this volume primarily concerns sites within the Jumanos Division. Redrawn from a National Park Service map (Ivey 1988:figure1) and from maps in Mera (1940).

formed where precipitation was retained for longer periods of time on buried substrate, or where the water table was unusually high. With continued climatic warming, some of these playa lakes dried up entirely and are hardly noticeable on the modern landscape. In other places, lakeshore features such as beach sand deposits, cliffs, terraces, and backwater ridges record the location of the edges of the ancient lakes. Within the lake basins, wind erosion and deflation created a gently rolling landscape of dunes made of the silt and clay from the dry lakebeds (Bourlier 1970; Allen 2005).

The largest salt lake in the Estancia Basin, Laguna del Perro (Lake of the Dogs), is one remnant of ancient Lake Willard. Laguna del Perro currently is about 19 kilometers long and about 1.6 kilometers wide, but the depth of the water is rarely more than thirty centimeters. Brilliant white deposits of gypsum and halite mark the changing location of its shoreline over time, and the thin layer of water shimmers brightly in the sunshine. In the Estancia Basin, these crystalline mineral deposits consist mainly of sodium chloride, or common table salt. The salt deposits are actually very thick and quite extensive; they were mined as a source of dietary salt and also as a preservative in the days before refrigeration. People from various pueblos along the Rio Grande made excursions to the Estancia Basin to collect salt to take home and to use in trade (Colton 1941; Lange 1959). When the Spanish arrived, they too recognized the value of salt for flavoring and preservation. In addition, they were most interested in the industrial use of salt in processing silver ore. Salt therefore became one of the major items of trade between this area and the Spanish occupation in Mexico. The salt was packaged in leather bags and exported to Chihuahua, over 1,000 kilometers away, along with other exports such as Indian captives (Gregg [1844]1967; Works 1992). Years later, Spanish- and English-speaking settlers in the area also exploited the salt deposits for their own use and profit (Hayes et al. 1981).

Because the natural drainage of the Estancia Basin leads to these interior lake basins, there are no permanent streams (Meinzer 1911), and water for household and industrial use has always been in short supply. In the northern portion of the Salinas area, ancient and modern occupants obtained drinking water from several snow-fed springs that emerge along the eastern slopes of the Manzano Mountains. Small springs are found even today near the Spanish period pueblos of Abó and Quaraí. Throughout the rest of the Salinas area, however, the only surface water is the small amount of rainwater retained for a short time in several tiny playa lakes that are scattered across the low-lying plains. In the last century, some ranchers have tried to make some of the playa lakes bigger and deeper in order to capture and retain more of the summer rainfall.

In the Estancia Basin itself, deep wells provide fresh water for irrigation agriculture today (New Mexico Interstate Stream Commission 1974; Yonts and Nuland 1984; White 1994). The underlying water-bearing sedimentary rock slopes down to the south, however, making commercial irrigation less feasible there, even with modern drilling technology. People in the southern portion of the Salinas area therefore relied, and still rely, on water from artificially dug cisterns and shallow underground wells. These shallow wells do not reach into the deeply buried sedimentary rock of the water table itself, but tap into a layer of less permeable silt and caliche (deposits of calcium carbonate) that underlies the loose surface sand. This layer slows the percolation of rainwater down through the sediments and retains moisture, forming a so-called perched water table. This shallow water source is not nearly as dependable, productive, or laterally extensive as a proper water table, and the depth to water is highly variable across space, depending on the topography and continuity of this less permeable subsurface layer. Nevertheless, these small playa lakes and shallow subsurface wells were both important sources of household water in the ancient and also more recent past.

Spanish records describe the use of thirty-two artificial wells that tapped into this near-surface water around Gran Quivira Pueblo (Sholes 1940:282; Toulouse 1945; Howard 1959; see also Evans 1951 and Scarborough 1988). Later explorers could not relocate any proper wells, however, and the apparent lack of readily available water near Gran Quivira has inspired much comment about the ingenuity and hardiness of the ancient population. For example, one Spanish commentator, Nicolas de Aguilar, marveled, "their lack of water is so acute that they are accustomed to preserve their urine to moisten the earth to make walls" (Hackett 1937:142).

Historic records from the homesteading period show one relatively more permanent pond (a playa lake) about a mile west of Gran Quivira Pueblo. This playa apparently provided a small but fairly constant water supply from rainfall, runoff, and also from seepage up from the subsurface perched water layer. However, the nearest dependable spring near Gran Quivira reportedly was located at the base of the Gallinas Mountains to the east, about twenty-four kilometers away. According to some accounts (Gregg [1844] 1967), water from this spring ran to the pueblo within some sort of man-made ditch or canal, but even in the 1800s, explorers could find no sign of any water supply system. It is hard to imagine how an open ditch could transport sufficient water for a town over this distance at such a low slope, and various researchers have argued that these practical engineering problems make it highly unlikely that any ditch ever could have existed (see discussion in Toulouse 1945).

The use of shallow, hand-dug wells, however, is well established throughout the Salinas region in the historic as well as modern period. Again, these shallow wells did not extend to the (sandstone) water table, but tapped into a stratum of water-saturated sand and underlying silt and caliche, where rainwater slowed as it trickled through from the surface. Water retention features such as cisterns supplemented the water supply in both ancient as well as modern times (e.g., Toulouse 1945, and a detailed discussion in Bandelier [1881]2008:27–30; see also Dominguez and Kolm 2005 for a more general discussion of ancient water management).

In general, however, this shallow water in the Salinas area has always been considered rather bad tasting and "gyppy," that is, full of gypsum, mostly suitable for livestock and of insufficient quantity for irrigation (Bates et al. 1947). Despite these problems, the area sustained a large population of Pueblo Indians at the time of Spanish contact in the late 1500s, who made a living by dry farming maize, beans, squash, and also cotton. Later homesteader settlers also managed to find enough water for household needs (Culbert 1941; see also Link and Link 1999), even if they did not have access to water for irrigation (see Mimms and Zaumeyer 1947). In fact, local ranchers today reminisce about the early half of the twentieth century when they relied solely on these shallow household wells for drinking, watering small kitchen gardens, and washing (Helen Towell, personal communication 1992; Louise Kite, personal communication, 2007).

The Spanish Period Pueblos

Among the pueblos of this nation there is a large one which must have three thousand souls; it is called Xumanas, because this nation often comes there to trade and barter

—Fray Alonso de Benavides, *Fray Alonso de Benavides' Revised Memorial of 1634*

Although sparsely populated today, the Salinas region was home to thousands of people when the Spanish arrived in the late 1500s (Berman 1979; Hammond and Rey 1940; Spicer [1961]1981; Barrett 2002). Although

Hayes (1993) notes that Fray Alonzo de Benavides may have overestimated the population of Gran Quivira itself, it was certainly the largest of the Salinas pueblos and one of the larger pueblos in New Mexico at that time (Hayes 1993:11). The pueblos of Quaraí and Abó were also quite large, with an estimated population of about six hundred and eight hundred inhabitants, respectively, at the time of Spanish contact (Prince 1915:338, 349; see also Noble 1993).

Based on Spanish records and the known distribution of linguistic groups among the historically occupied pueblos, H. P. Mera (1933) divided the Salinas Province into three "divisions" or cultural subregions: the East Tiwa, Tompiro, and Jumanos Divisions. The circled areas in figure 2 show the approximate areas of each subregion of the Salinas Province.

Mera's "East Tiwa Division" is located in the northern portion of the area. Here, the Spanish encountered Tiwa-speaking people in the pueblo of Quaraí and also in a number of small pueblos along the eastern foothills of the Manzano Mountains. The original Indian inhabitants left these pueblos in the late seventeenth century, a few years before the Pueblo Revolt of 1680. After Spanish reconquest of the area in 1692, Spanish-speaking settlers moved north from Mexico and established towns on Spanish land grants on or near the earlier Indian communities. The modern towns of Tajique and Chililí mark the location of older pueblos (Mera 1940:22), but today Tiwa-speaking people live in the Rio Grande valley. Southern Tiwa is used in the modern pueblos of Isleta and Sandia, and Northern Tiwa is spoken at the pueblos of Taos and Picuris (Cordell and McBrinn 2012).

Other Indian groups in the Salinas Province spoke a different language, Tompiro, which was closely related to the Piro language spoken by Indians living in the Tularosa Basin and the southern reaches of the Rio Grande, the Rio Abajo. The Spanish administrators recognized these two major language differences in the Salinas Province and also noted other political and possibly cultural differences among the several Tompiro-speaking pueblos. In particular, they separated out Abó Pueblo as different and grouped together the southern Tompiro pueblos near Gran Quivira with those located farther east, along the slopes of Jumanos Mesa. Mera (1940) followed the Spanish distinction between the northern and southern Tompiro-speaking pueblos of Salinas, identifying Abó and Ténabo as "Tompiro Division" pueblos, and the southern Tompiro pueblos as the "Jumanos Division" pueblos (see figure 2). The Jumanos pueblos thus include Gran Quivira and its neighbor Pueblo Pardo, and also Pueblo

Blanco (Tabirá) and its neighbor Pueblo Colorado. Among all these large late pueblos, however, only Quaraí, Abó, Gran Quivira, and Pueblo Blanco show significant Spanish occupation.

The relationship between the Jumanos pueblos and their Plains-dwelling neighbors is ambiguous. The Jumanos Indians were early bison-hunting nomadic traders of the southern Plains. After the seventeenth century, hunters related to the modern Apache moved into the area from the northern plains of Colorado, apparently replacing the Plains Jumanos in this area (Hayes 1993; Hickerson 1994; Brooks 2002). The fate of the original Jumanos of the southern Plains is not clear (Hickerson 1994).

Spanish documents explain that the Jumanos Indians were *rayado*, or "striped," referring to some form of painted or tattooed body decoration. The Jumanos pueblos described by the Spanish are thus not necessarily those pueblos where the Jumanos people lived, but where they customarily came to trade. This identification of the Jumanos pueblos as the trading partners of the Jumanos of the Plains is substantiated by study of the human skeletal material from the 1970s excavations at Gran Quivira Pueblo (Hayes et al. 1981). The skeletal analyses show that the people were biologically most similar to other Pueblo Indian groups and unlikely to be of Plains origin (Turner 1981; Peeples in press). However, the exact nature of the biological and social relationships between Plains and Pueblo Indian groups such as those in the Salinas region is likely much more complex, involving variable degrees of captive-taking, slavery, intermarriage, intergroup adoption, and other exchanges of adult men and women and also children (Brooks 2002:18).

The Spanish Mission Churches and Pueblos

With the wealth of historic documents and imposing standing architecture of the Franciscan Catholic churches at Abó, Quaraí, and Gran Quivira (Ivey 1988), it is not surprising that the Spanish period missions and their pueblos have been the most intensively studied archaeological sites in the area. Salinas Pueblo Missions National Monument is one of several archaeological monuments in New Mexico, and tourism to these pueblos and their Catholic missions plays a large role in the current economy and culture of the area. The sites themselves also play a large role in local history, including various popular renditions of colorful, if not always accurate, legends left over from the Spanish occupation. These stories evoke the grisly tortures of the Holy Inquisition, popular stereotypes of marauding Apaches from the Plains, and romantic tales of buried treasure. An early writer from the 1920s, Earle R. Forrest, explains:

> No other spot in the entire United States, with the exception of the Seven Cities of Cibola, can equal with romantic interest the name of Gran Quivira; and like the Seven Cities, it was only a myth invented by the Pueblos to lure Coronado away with the hope that the Spaniards would perish on the Great Plains . . . Tales of gold buried to keep it from Coronado have lured many adventurer down to our own times to search for the hidden treasure of Gran Quivira; and to this day the charm of the mysterious hovers about the Pueblos (1929: 147–148).

At some point, however, the imaginary La Gran Quivira became identified with the historic site of Pueblo de las Humanas, Gran Quivira Pueblo. One story from the 1880s reported that "1,600 burro loads of gold and silver" could be found there, along with the "golden bells" left by the Spanish priests (Cheney 1934:39). As one might expect, these stories of the riches of Gran Quivira form an enduring background to local history (Cheney 1934; Vivian 1979:31; Beckett 1981). This theme of Gran Quivira's undiscovered treasure is featured in older and also more recent novels (Corbyn 1904; Casey 2007) and in more ephemeral online blogs, travel guides, and other websites.

Part of the continuing appeal of such legends perhaps stems from our intuitive recognition that the idea of the Spanish burying their valuables and beating a hasty retreat is not an unlikely result of military occupation. As any conquering group has discovered, it is one thing to establish political control of an area, and another thing altogether to maintain that control on a long-term basis. Indeed, the Spanish were constantly dealing with innumerable resistance movements up until the time they were defeated and forced out of New Mexico.

By the time of the Pueblo Revolt in 1680, however, the Spanish mission priests and the Salinas Indians had already left the area (Schroeder 1979; Ivey 1988; Noble 1993; Kessell 2008). Drought and famine, raids from the neighboring Apache from the Plains, and the hardships and diseases that accompanied Spanish occupation took an enormous toll in human suffering and death. Spanish records indicate that most of the survivors from the Tiwa-speaking pueblo of Quaraí moved north to join other Tiwa pueblos along the east side of the Manzanos Mountains. The inhabitants of the Tompiro-speaking pueblos, Abó, Gran Quivira, and Pueblo

Blanco, generally moved west to seek refuge among the pueblos along the southern Rio Grande. Today, people who live in the Indian pueblos of Isleta and Sandia trace their ancestry to the Salinas pueblos. Other refugees moved south along the Rio Grande, toward Las Cruces, where they formed a community that is now called Isleta del Sur (Parsons 1975; Hayes 1981; Hayes et al. 1981; Murphy 1993a, 1993b; Noble 1993).

Some Indians did return briefly to reestablish the pueblo of Tabirá (Pueblo Blanco; Squires 1923), and excavations in 1999–2000 by Arizona State University archaeologists did identify some late room construction that may stem from this reoccupation (Graves and Spielmann 2000). However, neither the Spanish nor the Indian populations were able to sustain any lasting reoccupation of the area (Ivey 1988).

Centuries later, in the early 1800s, the Spanish government granted a large territory in the Estancia Basin to Col. Bartolome Baca, but later territorial claims by Mexico led to further subdivision and land allocations under Mexican law. After New Mexico became part of the United States in 1846, English-speaking homesteaders from Texas tried subsistence dry farming. They were followed by an influx of Spanish-speaking settlers from Mexico in the 1920s who tended to focus on sheepherding (Herrman 2003:15–23). In the early parts of the twentieth century, dry farming of pinto beans and other dry beans (Mimms and Zaumeyer 1947) briefly made Mountainair the "Pinto Bean Capital of the World."

Pinto bean farming supported hundreds of people in many small towns throughout the area. For example, the historic town of Gran Quivira (near the archaeological site) was home to some six hundred people in the early years of the twentieth century (Link and Link 1999; Herrman 2003). The combination of soil erosion, drought, and economic changes during and after World War II contributed to the end of the pinto bean era (Rautman 1994). The regional economy today relies on cattle ranching and also archaeological tourism, with Mountainair representing the "Gateway to Ancient Cities" and the base of operations for Salinas Pueblo Missions National Monument (Rautman 2008a).

The mission churches and pueblos of Abó, Quaraí, and Gran Quivira thus continue to play a large role in local culture. Many local residents work for the National Park Service, the Forest Service, or in various related services related to the tourism industry, and are quite knowledgeable about and concerned with their local history. Many of the ranching families in this region have lived here since homesteading days, and their family memories are deeply intertwined with the archaeological sites. One sees this intimate connection through the life of Park Service ranger Fredrico

Sisneros, whose family long held title to the ruins of Abó. In 1938, he donated this land to the state of New Mexico to form a state monument (Ivey 1988). In 1944, he was appointed caretaker and later, ranger, a position he held until his death in his nineties (*New York Times* 1988).

His grave is located just outside the ruins of the mission church of San Gregorio de Abó, where the dusky red sandstone masonry of the church contrasts sharply with lush green grass around a spring. A shallow arroyo drains into Abó Pass between the Los Pinos Mountains to the south and the Manzanos to the north. Geology students come to this area to pay homage to the type location of the distinctive red sandstone of the Abó formation (Lee 1909), which is visible along the highway from Belen to Mountainair. Students of archaeology come to study the many examples of rock art pecked into the smooth red sandstone, and tourists come to visit the ruins.

Quaraí is somewhat similar in appearance and setting, tucked into a small verdant valley by a spring; tourists linger in the shade of the large cottonwood trees and picnic by the Spanish period irrigation ditch. Fruiting bushes of chokecherry, currants, and gooseberries are likely descendants of plants that supplied the pueblo before the Spanish came (Hill 1998; Morrow 2006; see also Anschuetz 2006); their presence adds a note of intimate connection with the ancient inhabitants. Quaraí has a rather special status among the Salinas pueblos; this was the seat of the Holy Office of the Inquisition in the Jurisdiction of Salinas. One of the last subjects of the Inquisition, Bernard Gruber, was held for a number of years in prison at Quaraí; the story of his short-lived escape is recorded by the rather grisly place-name of the region south of Salinas: the Jornada del Muerto, or Dead Man's Journey (Sanchez 1993; Wilson 1993; Kessell 2008).

South of Mountainair, the pueblo of Gran Quivira presides over a more dramatic setting. This pueblo was built atop a small rocky ridge east of Chupadera Mesa, where it commands a view of the Medanos Plains below. Its handsome light gray limestone and upland setting make it visible for miles away, and the more arid landscape gives it a more imposing presence amid the scattered piñon and juniper trees, snakeweed, and sage.

Partly as a result of this history of settlement, and also due to the continued importance of the National Park Service and its employees in supporting archaeological research, most of the archaeological survey and excavation over the last hundred years has occurred in and around these three mission sites. Table 1 is a list of the major archaeological projects in the Salinas region, focusing primarily on research that led to scholarly publication. The references are not meant to be an exhaustive list; instead,

Table 1. Major archaeological research projects in the Salinas Province*

Fieldwork Date	*Site*	*Principal Investigator and Institutional Affiliation*	*Primary References*
1882	Quaraí–mapping (LA 95)	Adolph F. Bandelier (Archaeological Institute of America)	Bandelier 1881, 1890; Lange and Riley 1966:388
1916–1917	Gran Quivira (LA 120) ("treasure hunting")	Jacobo Yrisarri (No institutional affiliation; fined under the 1916 Antiquities Act)	Vivian 1979
1913, 1916, 1934–1936, 1939–1940, 1944	Quaraí (LA 95)	Edgar L. Hewett (University of New Mexico); Donovan Senter, Albert Ely, and Ele M. Baker (School of American Research and the Museum of New Mexico); Wesley R. Hurt (Works Progress Administration)	Senter 1934; Ely 1935; Baker 1936; Reed 1940; Ivey 1988:9–21; Hurt 1990
1923	Gran Quivira, "the only known survey of the entire mound area of Gran Quivira"	Ida Belle Squires and Anna Shepard (Teacher's College, San Diego, CA)	Squires 1923
1923, 1925	San Buenaventura Church, Gran Quivira (LA 120)	Edgar L. Hewett (School of American Research and University of New Mexico)	Vivian 1979; Ivey 1988
1930, 1932, 1934	Gran Quivira (LA 120) ("permit to dig for treasure" issued in 1930)	J. B. Wofford, Alfred J. Otero, and Jacobo Yrisarri (No institutional affiliation)	Vivian 1979; Beckett 1981
1930s	Survey of Chupadera Mesa and surrounding area; 60 sites	H. P. Mera (Museum of New Mexico)	Mera 1940
1941	Pueblo Pardo (LA 83)	Joseph H. Toulouse (National Park Service);	Toulouse and Stephenson 1960

Table 1. (continued)

Fieldwork Date	*Site*	*Principal Investigator and Institutional Affiliation*	*Primary References*
		Robert L. Stephenson (Washington and Jefferson College and the University of Oregon)	
1944	Abó (LA 97)	Bertha Dutton (University of New Mexico)	Toulouse 1949
1951	Gran Quivira (LA 120)	Gordon Vivian (National Park Service)	Vivian 1979
1953	Excavations at LA 2579 (Pipeline site)	Franklin Fenenga (Texas Technological College)	Fenenga 1956
1954	Excavations at LA 2579 (Pipeline site)	Earl Green (Texas Technological College)	Green 1955
1957	Pueblo Blanco (Tabirá, LA 51)	Johnnie Cress (No institutional affiliation)	Cress 1957; Wilson et al. 1985
1959	Pueblo Blanco (Tabirá, LA 51)	Stanley Stubbs (Museum of New Mexico)	Stubbs 1959; Lovell 1963; Wilson et al. 1985
1964	Small sites near Gran Quivira	Douglas Scoville and Ronald Ice (National Park Service)	Ice 1968
1965, 1966, 1967	Gran Quivira Pueblo (LA 120)	Alden C. Hayes (National Park Service)	Hayes 1981; Hayes et al. 1981
1969	Survey of PaleoIndian and Archaic sites in Estancia Basin	Thomas Lyons (University of New Mexico)	Lyons 1969
1970s	Gran Quivira area survey	Thomas J. Caperton (National Park Service)	Caperton 1981
1975	Estancia Basin excavation at Tillery Springs	Thomas Lyons (University of New Mexico)	Lyons and Switzer 1975

Table 1. (continued)

Fieldwork Date	*Site*	*Principal Investigator and Institutional Affiliation*	*Primary References*
1981	Survey within the Gran Quivira property boundaries	Patrick H. Beckett (New Mexico State University and COAS Publishing and Research)	Beckett 1981
1982	Test excavations at Kite Pithouse Village	Regge N. Wiseman (Museum of New Mexico); Patrick H. Beckett (COAS Publishing and Research)	Beckett and Wiseman 1982; Wiseman 1986
1984, 1985, 1986	Gran Quivira (LA 120)	Katherine A. Spielmann (University of Iowa and Arizona State University)	Spielmann 1992
1980s	Ténabo (LA 200)	Stuart Baldwin (University of Calgary)	Baldwin 1983
1986	Chupadera Arroyo Survey	John Montgomery and Kathleen Bowman (Eastern New Mexico University)	Montgomery and Bowman 1989
1986	Kite Pithouse Village (LA 38448)	Alison E. Rautman (University of Michigan)	Rautman 1990
1988	Salinas Survey	Katherine A. Spielmann (Arizona State University)	Spielmann and Eshbaugh 1989
1989	Pueblo Colorado (LA 476)	Katherine A. Spielmann (Arizona State University)	Spielmann 1998a
late 1980s	Quaraí (LA 97)	Walter K. Wait and Peter J. McKenna (National Park Service)	Wait and McKenna 1990
1992, 1993	Quaraí (LA 97)	Katherine A. Spielmann (Arizona State University)	Spielmann 1993, 1994a
1992, 1993	Pueblo de la Mesa (LA 2091)	Alison E. Rautman (Michigan State University)	Rautman 1992, 1993a, 1995c
1994	Kite Pueblo (LA 199)	Alison E. Rautman (Michigan State University)	Rautman 1995a, 1995b; Rautman and Heyman 2011

Table 1. (continued)

Fieldwork Date	*Site*	*Principal Investigator and Institutional Affiliation*	*Primary References*
1999, 2000	Pueblo Blanco (Tabirá, LA 51)	Katherine A. Spielmann and William R. Graves (Arizona State University); Alison E. Rautman (Michigan State University)	Graves and Spielmann 1999, 2000a
2001, 2003	Survey of jacal sites and Glaze A pueblos	Matthew A. Chamberlin (Arizona State University)	Chamberlin 2001, 2008
2005	Abó (LA 97)	William R. Graves (University of Iowa)	Graves 2005
2007, 2008	Frank's Pueblo (LA 9032)	Alison E. Rautman (Michigan State University); Matthew A. Chamberlin (James Madison University, VA)	Rautman 2008b; Brouwer et al. 2009; Rautman 2013c
2009, 2010	Sites on Turkey Ridge and at Pueblo Seco (LA 9029)	Matthew A. Chamberlin and Julie Solometo (James Madison University, VA)	Personal communication

*The National Park Service has sponsored numerous small salvage excavations associated with stabilization work and construction of facilities for modern visitors at the mission pueblo sites of Abó, Quaraí, and Gran Quivira. The references included here represent only the basic archaeological field reports or monographs.

they are provided here as a source of entry into the scholarly literature. Other short archaeological and stabilization reports regarding activities on NPS property are on file at Salinas Pueblo Missions National Monument headquarters in Mountainair.

The Park Service funded various large-scale excavations at Abó, Quaraí, and Gran Quivira, research that was originally associated with developing these mission sites for tourism (e.g., Toulouse 1949; Vivian 1979; Hayes 1981; Hayes et al. 1981; Ivey 1988; Hurt 1990). In addition, the National Park Service has sponsored a great deal of smaller-scale archaeological work associated with park functions, including construction for modern tourist facilities and stabilization. More recently, a small group of university-based archaeologists have worked together and also separately to investigate the Salinas pueblos in more detail. Kàtherine Spielmann of Arizona State University first began a long-term project at the several large Spanish mission sites in the 1980s, comparing economic and political organization before and after Spanish occupation. William Graves continued some of that work at Pueblo Blanco and also Abó (Graves and Eckert 1998; Graves and Spielmann 1999, 2000a; Graves 2002).

Early Village Sites of the Salinas Province

My study of village sites in the Salinas Province, however, concerns time periods that are much earlier and sites that are much smaller than the Spanish period pueblos. In this volume, I describe my research at four different sites within the Jumanos District, each of which seemed, at least at first glance, to be a village site, with evidence of multiple residential structures clustered within a limited locale, and a great many artifacts visible on the ground surface. The locations of the four village sites are shown in figure 3.

After working at Gran Quivira with Katherine Spielmann when I was a graduate student (see Spielmann 1992), I began this research by working on an early pithouse site in 1986, Kite Pithouse Village, and continued with excavations at three different early pueblos: Kite Pueblo, Pueblo de la Mesa, and most recently Frank's Pueblo. In the last few years, Matthew Chamberlin and Julie Solometo have conducted surveys and also some excavations at various early masonry pueblos near Turkey Ridge and further west. Other researchers such as Tiffany Clark have investigated different specific topics in this area, focusing on the analysis of artifacts derived from these and other excavations (e.g., Clark 2006). However, the reality of the situation is that the archaeological record of early village sites in Salinas is not well known.

For example, despite a couple of intensive surveys within the boundaries of the Gran Quivira property (Beckett 1981), and Spielmann's survey of the Medanos Plains near Gran Quivira (Spielmann and Eschbaugh 1989), we simply do not have even basic information about the number and size of the earliest village sites (the pithouse sites) more generally across the area. Subsequent Jacal period sites are similarly poorly known, although Chamberlin's (2008) survey on Chupadera Mesa southwest of Gran Quivira is a major contribution.

The pueblo sites with masonry walls are more visible on the landscape, but many of those sites on private land have not been revisited since the

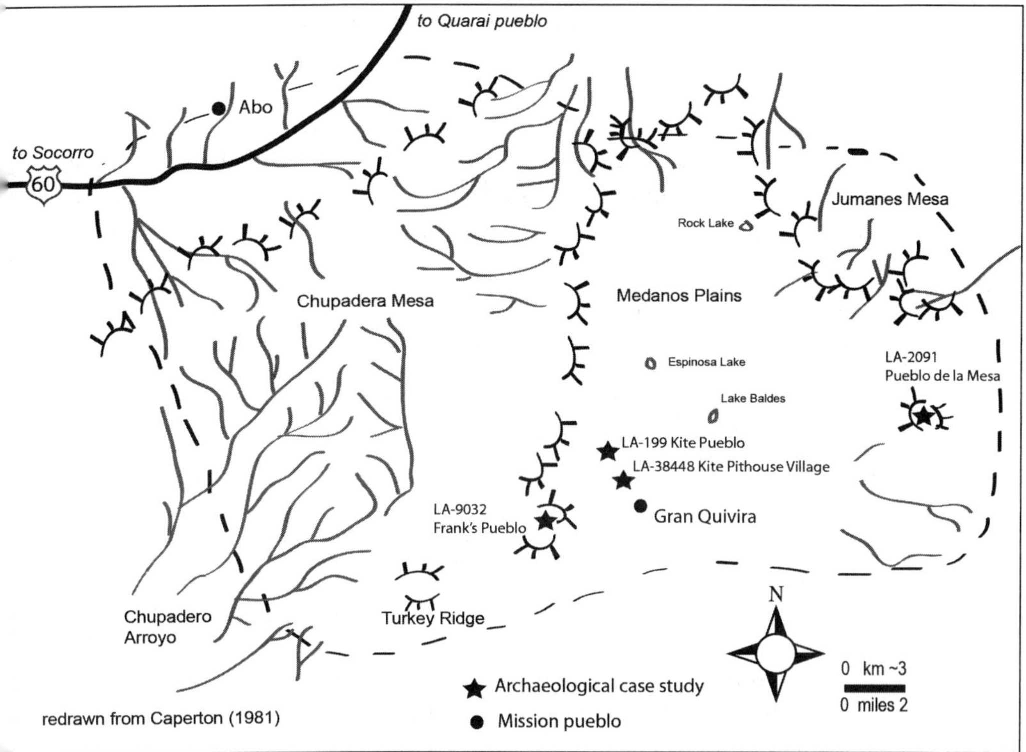

Figure 3. Map of the study area near Gran Quivira Pueblo. The stars show the location of the four archaeological sites discussed in detail in this volume. Modified from Caperton (1981:xii).

archaeologist H. P. Mera mapped them in the 1930s and 1940s. John Montgomery and Kathleen Bowman (1989) remapped some of these pueblos on the west side of Chupadera Mesa (see also Kyte 1988), an area that is adjacent to the Salinas Province. More recently, Ward Beers (2012) investigated the spatial relationships of known pueblo sites in the Jumanos cluster and their intervisibility, a project that involved revisiting some large pueblos and some archaeological survey in targeted areas between them. As a result of these studies, we do have fairly good information on the locations and spatial organization of the more visible masonry sites. However, we still know very little about community organization within each village site.

At Abó, Quaraí, and Gran Quivira, more intensive survey and salvage excavation has occurred, but this effort concerns primarily the later time periods, and is restricted within the park boundaries. In sum, except for the extensive Park Service excavations of one of the large house mounds (Mound 7) at Gran Quivira Pueblo in the 1960s and 1970s, the extent of archaeological information available for the different early village sites is pitifully small compared to some well-known areas in the Southwest where generations of field school students have recovered detailed chronological information from many sites and structures over decades of study.

Why Study Early Salinas Villages?

Given this paucity of research, why is the Salinas region appropriate for studying such an important worldwide phenomenon as early village society? The advantages of the Salinas area encompass both highly pragmatic and also significant theoretical issues. First, the research reported here derives from the same local area. Although the locales and specific settings of the sites vary, the fact that they all are located fairly close to one another has the effect of holding the variable of "site location" more or less constant, at least within the scale of the region. In addition, all of the sites considered here fit within H. P. Mera's (1940) "Jumanos cluster." As a result, my research projects described here provide a unique data set that allows us to evaluate evidence of village life and the social changes over a span of time that encompasses early village societies within a single defined area.

Another quite pragmatic advantage highlighted here is that each of the sites described was excavated using the same general methods and procedures for screening, sampling, and basic analysis. Of course, there was a certain learning curve involved, and I often wish I might have approached my early projects with whatever technology and wisdom that I gained later in time. However, in fact I did try to maintain a degree of consistency in recording sites, features, and artifacts for just this reason. Even though the original research objectives for each project varied, and although not all sites yielded the same kinds of data, the data sets that were recovered are in fact comparable in terms of sampling strategy and methods of recovery, identification, and laboratory treatment. These data—the plant and animal remains, the debris from stone tool manufacturing and use, and the pottery types recovered—provide a general sense of everyday life at each early village site, although the degree of detail of different aspects of life varies depending on the specific samples recovered.

For the sites with aboveground architecture, excavations also provide a detailed history of each village's construction, remodeling, and use. For these sites, much of the information available about village community dynamics will come from evidence of the built environment and evidence of how people modified these constructions to fit with changing social dynamics over time (e.g., Bradley 1993; Crown and Kohler 1994). However, at all these sites, the architecture is not the only source of information. There are also excavation data that provide additional evidence regarding room function, people's everyday activities, and the process of site abandonment.

Other advantages of the Salinas Province data set are related to general issues in the archaeology of the Southwest. For example, in some areas of the Southwest, pueblo sites record repeated in-migrations and population movement (e.g., Bernardini 1998, 2005; Spielmann 1998; Duff 2005; Cordell et al. 2007). In contrast, as far as we can tell, the Salinas Province apparently experienced comparatively greater demographic continuity across much of its history (see Spielmann [1996], whose observations have been reinforced by later work, including Rautman [2000] and Chamberlin [2008]). Of course, people in the Salinas region were not confined to their pueblos or their immediate surroundings. They moved around the landscape to access resources such as salt and clay, they maintained diverse and distant social and economic ties with other groups, and they exchanged ideas as well as pottery and other objects with many different groups across the Southwest and southern Plains. However, it is also true that this area seems to have been less impacted by the repeated, frequent, and relatively large-scale population movements that we see elsewhere in the Pueblo Southwest.

In the Salinas Province, much of the population movement that we see throughout the archaeological record—for example, continued population aggregation into fewer and larger pueblo sites such as Gran Quivira, Abó, and Quaraí—is probably best interpreted as a population reorganization within the Salinas region as a whole (Spielmann 1996; Rautman 2000). Some caveats to this generalization are discussed in more detail in chapter 6, especially for the later pueblos, but in general, the Salinas region provides an interesting counterexample to the experience of village dwellers in some other regions within the Pueblo Southwest.

Another major advantage of this Salinas area sample stems from the fact that all of the early village sites occur relatively late in the archaeological sequence of the Southwest. Table 2 compares the local chronology with the more familiar Pecos terminology of Basketmaker and Pueblo time periods. In other areas of the Southwest, issues of village origins and dynamics are closely intertwined with issues of the timing of maize domestication and adoption, agricultural dependence, and sedentism (Cordell

Table 2. Time periods and significant historic events in the Salinas Province

Salinas Chronology	*Approximate Dates in the Salinas Area**	*Corresponding Time Period in the Pecos Chronology*
Paleoindian period	Clovis 9500 to 9000 B.C.; Folsom 8800 to 8300 B.C.; Plano 8300 to 6000 B.C. (Tainter and Levine 1987:12–26)	
Archaic period	About 6000 B.C. to A.D. 600 (Tainter and Levine 1987:26–31)	Basketmaker II (about 500 B.C to A.D. 500)
Early Pithouse period	A.D. 600 to 900	Basketmaker III (about A.D. 500 to 750); Pueblo I (about A.D. 750 to 900)
Late Pithouse period	A.D. 900 to 1150 (or possibly 1200 at the latest; Beckett 1981)	Pueblo II (about A.D. 900 to 1150)
Jacal period	A.D. 1100 to 1300; dates may overlap with the latest pithouse occupation	Pueblo II and III
Early Pueblo period (adobe pueblos)	A.D. 1150 to 1300; dates are uncertain; may overlap with some jacal use; sites have few if any Glaze A sherds	Pueblo III (about A.D. 1150 to 1300)
Early Pueblo period (early masonry pueblos)	A.D. 1313 to 1410 (dated by the presence of Glaze A pottery types and the absence of subsequent glaze types)	
Late Pueblo period	A.D. 1410 to 1680	Pueblo IV (about A.D. 1300 to 1600)
Spanish contact and conquest	A.D. 1539 and 1540: the first Spanish expeditions in New Mexico; A.D. 1582 to 1583: the Espejo expedition visits the Salinas region (Vivian 1979:12); October 1598: the leaders of the Salinas pueblos sign don Juan de Oñate's Act of Obedience to the King of Spain (Hayes et al. 1981:4; Murphy 1993:22)	Pueblo IV and V

Table 2. (continued)

Salinas Chronology	*Approximate Dates in the Salinas Area**	*Corresponding Time Period in the Pecos Chronology*
Spanish missions established	A.D. 1622: the first Franciscan missionary priests arrive in Salinas (Hayes et al. 1981:4; Murphy 1993:26)	
Pueblos and missions abandoned	1671: Gran Quivira abandoned; 1673: Abó abandoned; 1677: Quaraí abandoned (Hayes et al. 1981:4–6; Murphy 1993:59–60)	Pueblo V (after A.D. 1600)
Spanish land grants	Early 1800s (Herrman 2003)	Pueblo V
American homesteading era	The late 1800s, with most immigrants arriving after 1900 (Herrman 2003; Link and Link 1999)	
Preservation of the Salinas pueblo mission sites	1909: Gran Quivira National Monument established; 1930s: Quaraí and Abó come under public ownership; 1980: Salinas National Monument established; the name later changed to Salinas Pueblo Missions National Monument (Murphy 1993:63)	

*Dates of the time periods in the Salinas region are based on Caperton (1981) unless otherwise noted. The dates associated with the periods of the Pecos classification system are from Cordell and McBrinn (2012:74). The dates for glaze ware pottery are provided in table 14 in this volume.

and McBrinn 2012). In contrast, in the Salinas region, the earliest known pithouse village sites date to about A.D. 600, long after maize was domesticated and productive varieties were widely adopted across the Southwest (see Staller et al. 2006). At the time of early village sites in the Salinas region, the basic Native American economic pattern was already well established, dominated by the "Three Sisters" of maize, beans, and squash (see Mt. Pleasant 2006).

In this case study, therefore, several of the interconnected strands of evidence that complicate the study of early village origins are already untangled. The earliest pithouse-dwelling Salinas villagers knew about cultivating domesticated maize, and the abundant remains of maize cobs and kernels at all four sites attest to its continued economic importance across time. But how did maize agriculture fit into the overall economic strategies of the Salinas people? In many areas of the world, we expect to see a trend toward increased dependence on cultivated foods. In many regions of the Southwest, there is also a widespread architectural change from pithouse to pueblo associated with this trend, in which increased local population densities and sedentism generally accompany an increased reliance on cultivated foods (e.g., Gilman 1987; Wills 1988a, 1988b; Kantner 2004; Doolittle and Mabry 2006, and more generally, Gurven et al. 2010).

In contrast, in Salinas, people's diets remained relatively constant over several centuries. By A.D. 900, pithouse-dwelling farmers in this area were already depending on maize agriculture for much of their diet. In this case, isotopic analysis of the composition of human bone recovered from Kite Pithouse Village (dating from about A.D. 900 to 1150) and from a later pueblo site, Pueblo de la Mesa, show that there was in fact very little overall dietary change over time (Katzenberg and Kelley 1991; Colyer 1996).

Within the Salinas area, therefore, the four sites reported in this case study represent a unique opportunity to study change over time in some aspects of village community dynamics while holding other common variables such as basic economy and dietary composition constant. Within one geographic area, we see marked differences over time in site architecture as pithouses are replaced by aboveground surface structures that we think of as pueblos. Over time, there is also a rather dramatic but also systematic region-wide change in construction methods. Early aboveground structures built of wattle and daub, locally called jacals, apparently overlap in their time of use with the latest occupation of some pit structures. As a building strategy, however, aboveground jacal architecture proved to be a successful concept, and jacal villages generally post-date the pithouse sites. A subsequent change to plaza-oriented adobe pueblos represents a new concept in village site organization, but again, use of some jacal sites and structures may overlap in time with the beginnings of adobe and (later) masonry pueblos.

As a result, the similarities of geographic setting and basic economy among all four sites provide a relatively stable background against which we can evaluate evidence of change in the social dynamics of early village communities in this particular region. My excavations over the last twenty years or so provide detailed information regarding the construction, spatial organization, and history of occupation of each of three pueblo village sites. While we lack a comparably detailed outline of site layout and occupational history for the earlier pithouse sites, the information available does establish some general parameters regarding structure size and construction, and changes in these variables over time.

Goals of This Study

This volume thus has a twofold purpose. First, the research presented here contributes to a larger discussion regarding how archaeologists interpret early village sites and early village society. Here, I do not view early village society as a particular temporal stage or a social type in the grand scheme of human evolution (compare, for example, to Otterbein 2009:77–82); rather, I am more interested in how certain early village communities "worked" or operated as living organizations within a world of similar small-scale early village societies.

The data presented here focus on information that we can infer based on the architectural evidence of the built environment: the evidence of village construction, remodeling, and use. However, while I consider specific architectural attributes as part of the study, I am not engaging in a formal space syntax analysis (cf. Hillier and Hanson 1984; Liebmann et al. 2005) or a detailed study of each site's architectural features per se (cf. Riggs 2001). Rather, I use evidence from the built environment as one important line of evidence, together with variable amounts of information from artifacts and also the remains of plants and animals at each site, to establish the ways in which each site's inhabitants experienced and also constructed the experience of living within a village.

My second goal for this volume is no less important, but is more geographically and topically localized to the archaeology of central New Mexico. Because my study area of the Salinas Province is not particularly well-known, even among Southwestern archaeologists, I also take some time in each chapter to situate my four specific case studies within the known archaeological record for the area as a whole, to articulate how my own research fits into larger debates, and to point out specifically where additional research is needed. My hope is that this volume will provide a synthesis of our current knowledge of the archaeology of early villages in the Salinas Province and will inspire another generation of archaeologists to work in this beautiful and fascinating region.

I will begin in the next chapter by discussing the local time periods represented by the four sites in more detail, and I will assess their potential for contributing to a more general model of village society dynamics and community operation within this local context. While other archaeologists have developed some thoughtful models for understanding the dynamics of village life elsewhere in the American Southwest, the archaeology of Salinas demonstrates that the range of variation in early village society is perhaps greater than we have imagined.

CHAPTER ONE

Village and Community

The dynamic social concept of village society used here draws on a long tradition of inquiry in anthropology and also archaeology. Different threads of this discussion in the archaeological literature weave back and forth, linking field observations and social models. In this chapter, I will review some of the relevant general concepts and explain how they will be used.

Archaeological models of early village communities derive from our general understanding of ethnographically observed small-scale society and also from theoretical models that draw on computer simulations, game theory, and other observations of interpersonal relationships and group behavior from studies in psychology, business, sociology, economics, and other social science fields. General models of early villages also rely on the known archaeological record of the earliest evidence for village sites, particularly those sites that date to the Neolithic period of the Near East and the Formative period in Mesoamerica—two areas where village sites are particularly ancient, and where archaeological research in the 1960s and 1970s has shaped many of our assumptions and models today.

I will then review some of the most common ideas that archaeologists use to understand early village societies in the American Southwest and will cite some comparative information from specific early village sites in the Mesa Verde region of southern Colorado. There are many important differences between the Salinas and Mesa Verde subregions of the Southwest, but these comparisons are helpful in providing contrast from an area of the Pueblo Southwest that is well-known for having sites that are widely considered to be early village communities, and also well-known in the sense of having been extensively studied. In the Mesa Verde area, large-scale and meticulous archaeological research began with the Dolores Archaeological Project in the 1970s and 1980s, and has been continued over time by the archaeologists of Crow Canyon Archaeological Center (in nearby Cortez, Colorado) and by several related multidisciplinary research projects through the Village Ecodynamics Project (see Kohler and Varien 2012 and references cited therein).

In this volume, I focus on the evidence of the built environment and related concepts that can help us best investigate the organization and operation of early village communities. The built environment includes the structures, features, spaces, and pathways that constitute the village that is visible to us today as an archaeological site, and which also figured prominently in the everyday life and activities of the people who once lived there (Rapoport 1982, 1988; Fisher 2009). In the case of the archaeology of the Salinas region, the built environment is sometimes the best evidence that we have for understanding social dynamics during particular time periods. When possible, however, I will bring in additional information from excavation, including the radiocarbon dates and analysis of plant and animal remains, of ceramic wares and stone tools, and of human diet. However, not all these categories of data are available for every site that I excavated.

This chapter includes a brief overview of the archaeological data that are available in the Salinas area for evaluating models of community organization across the different time periods and the aspects of the models that are most feasible for testing. Later chapters include details of research projects at different sites within the Salinas Province. Each chapter examines the details of community organization using the evidence observed from excavated village sites and that together provide the evidence for evaluating change over time.

Village Society, Village Sites, and Communities

In this study, the term village site refers only to a specific archaeological site where there is evidence of residential structures, and which the observing archaeologist thinks might represent a village. In contrast, the concept of early village society described by Bandy and Fox (2010) includes an inference of some widespread features of social life for the people who might have lived in such a village site. This particular expression of village life in and around the village site includes the people who live in the village itself, as well as any who might live elsewhere in field houses or in the surrounding countryside, but who participate regularly in social functions at the village site. The "institutions, practices, habits of thought, action, and expression that made village life possible" (Bandy and Fox 2010:16) are in fact those that we usually think of as forming a community.

Thus, these people and their interactions and social relationships at a specific time and place represent what I call the "village community." This definition recognizes that the community as a social group exists at a certain time and place, but that the community as a whole has an organizational lifespan that extends beyond any particular individual, and any particular individual's participation in the village community may vary (see essays in Canuto and Yaeger 2000; Varien and Potter 2008).

My goal here is to communicate the idea that the village site and the village community are not necessarily synonymous, but are closely related concepts. The village community involves social and political relationships but is also associated with a sense of place—but not necessarily associated with just one single locus where there are residential structures. Rather, the community as a whole exists physically within a cultural and natural landscape (Schlanger 1992; Amith 2005; Snead 2008), and also socially, culturally, and spiritually within history and imagination (Anderson 1991; Rogoff 2003; Christie 2009).

However, in many archaeological examples, as in my research, the logic begins with the archaeological site—that is, archaeologists hypothesize the existence of a particular village community from the material remains of the apparent village site itself. This hypothesis may be as simple as labeling a site a "village" because there is evidence of some structures—as I did when I began research as a graduate student at the site that I rather innocently called "Kite Pithouse Village." In fact, however, this identification of a village community should ideally involve a little more thought about whether the people who lived at the site really did exhibit any of the social relationships that we think of as constituting a community.

The concept of community is used in many different ways by archaeologists, anthropologists, sociologists, politicians, urban planners, community organizers, religious group leaders, and other people who concern themselves with "creating community" (e.g., Mattessich and Monsey 1997). Despite this variation, Southwestern archaeologists do seem to agree on some basic ideas (see Gilpin 2003). Communities are broadly defined as social groups of interacting individuals (see Murdock 1949), but these groups may be of variable size, scale, and composition (e.g., Adler 1989, 2002a, 2002b, and 2002c; Wilshusen 1991; Kolb and Snead 1997; Gilpin

2003). Communities need not always be manifest by a single archaeological site or locale (Mahoney et al. 2000; Adler 2002c), or in fact have any one defined spatial location, as we know from modern cyber communities. When communities do have a focal village site, there is also variation in settlement pattern and land use (e.g., Agorsah 1988). The village site may consist of one highly nucleated settlement, or be more dispersed across the landscape (Wills et al. 2012).

In the Southwest, for example, compact multistoried pueblos are considered to be highly nucleated village sites, while Euro-American farmers and ranchers as well as Navajo pastoralists are usually cited as illustrations of the more dispersed type of settlement pattern. Thus, the general concept of community, while conveying something regarding interpersonal connections, does not necessarily manifest in any particular architectural form or settlement. Conversely, an area that researchers identify as a pueblo does not necessarily represent a single community.

Among Southwestern archaeologists, recent studies have focused on just this issue: given a specific pueblo, can we really in fact assume that the pueblo's inhabitants represent a community? To what extent can modern or historic ethnographic information about relationships within and between pueblos inform us about the past (see also Isbell 2000)? Recent research among the Western Pueblo Indians highlights some of this known variation. For example, Duff (2002) and Bernardini (1998, 2005) have pointed out that repeated migrations among the Western Pueblo Indians resulted in pueblo villages that were composed of diverse groups that maintained significant social distinctions over time and competed with one another for political dominance and social status within the pueblo. They found that pueblo villages can be fairly loosely integrated, housing an assortment of different corporate groups (clans or subclans) that may have originated in different areas and have limited social ties or commitments to the pueblo as a whole.

There may be a number of distinctions apparent in the social roles of early-arriving groups compared to latecomers: the social situation at Grasshopper Pueblo in Arizona is one archaeological example (Lowell 2007; Riggs 2001, 2007). Fowles et al. (2007) argues for a somewhat similar situation among the inhabitants of the modern pueblo of Taos and also at the nearby archaeological site of Pot Creek Pueblo, New Mexico. Some modern pueblos, in fact, may represent fairly late settlements that were formed or reconstituted after the Pueblo Revolt (Liebmann et al. 2005). These conceptions of village politics emphasize the contingent and possibly temporary nature of community membership and village identity, and highlight the role of factions and group competition in social, economic, and political activities.

Pueblo Indian history provides several different examples of similar situations of intrapueblo differentiation, from the political importance of factional competition at Hopi (e.g., Whiteley 1988, 2008) to the case of the Hopi-Tewa at Hano (Dozier 1966). Historic records also show that ethnically different groups may together develop a new sense of identity, for example in the case of the modern pueblo of Laguna, which was founded in A.D. 1699 by members originating from various other pueblos (Prince 1915; see also Liebmann and Preucel 2007). As a result, we do know that archaeological and also historic or modern pueblos are quite variable in their organization as village communities. We know that village communities may fit easily with certain concepts of social identity, or they might incorporate diverse groups that differ in language and ethnic identity. Pueblo village communities can exhibit aspects of egalitarian social organization, and yet also exhibit significant social inequality (Smith 1983; Bender 1990; Brandt 1995; McGuire and Saitta 1996, 1998) and a variety of political leadership strategies (e.g., Mills 2000).

One archaeological concept related to community, however, is particularly useful for investigating early village sites, which are known to us primarily through the archaeological record. Kolb and Snead's (1997:611) definition helps bridge the gap between data (a given village site) and theory (a community) for the specific examples of small-scale village societies considered here in the American Southwest. They define a community as a "minimal, spatially defined locus of human activity that incorporates social reproduction, subsistence production, and self-identification." This kind of self-identification does not prevent individuals or groups from participating to a greater or lesser extent in multiple social communities, at least in theory (Isbell 2000; see also Yeager and Canuto 2000). Nor does it imply that one particular village site represents an entire community. On the other hand, in practice, some communities represent more formally organized and bounded groups, and members may find that participation in one group precludes membership in another. The nature of community composition and the ways in which boundaries are defined, reinforced, or transcended thus represent important concepts for investigation (e.g., Rogoff 2003; Lawrence 1996).

If we consider the concept of community in the sense of self-identification and group definition, we can also expect certain efforts toward social integration at a social scale larger than that of the nuclear family. Aspects of social integration may include large group activities and rituals, which in the Southwest may take place in specialized features of the built environment such as plazas or kivas (Hegmon 1989; Lipe 1989; Lipe and Hegmon 1989; Adler and Wilshusen 1990; Smith 1990; Adler 1993, 2002a). However, an often-overlooked aspect of group integration involves additional strategies of social differentiation, in which individuals establish and reinforce criteria for group membership by developing means of distinguishing members of one community from non-members. These concepts involve discussion of social differentiation, social boundaries, social signaling (e.g., Roscoe 2009), and/or issues of identity (e.g., Stark 1998; Stone 2000; Chamberlin 2001; Mills 2004b; Bernardini 2005; Riggs 2007; Eckert 2008; Varien and Potter 2008).

In this general conceptual model, actions and behaviors by individuals and social groups build and express community within the specific context of a social landscape (Snead and Preucel 1999; Snead 2008), often including participation in communal activities or events that may or may not exclude outsiders to the group, or by regular and repeated participation in activities that take place at a recognized focal area. The archaeological village site is the material representation of these repeated and patterned individual and group activities that constitute recurring participation in community life.

This concept of village communities allows us to link the archaeological data from the village site and surrounding regions to different expressions of the operation of a village community. The emphasis on activity and participation links people's individual experiences to abstract concepts such as the social group, and also recognizes gradations and variations in a given individual's intensity, frequency, and duration of participation in different groups. While there may be other kinds of group and individual interactions, only the most strongly patterned, repetitive, and pervasive are likely to be expressed physically in the built environment of the village site (McGuire and Schiffer 1983).

Identifying Communities in the Archaeological Record

Kolb and Snead's (1997) definition of community discussed above derives from their concern with identifying communities within a regional landscape. My primary interest here, however, begins with the material record at the scale of the individual site. When we as archaeologists identify a spatial cluster of structures or features in a given area, how can we determine whether this site was in fact a community?

This application of the concept of community is different from that used to describe how groups of people might engage in some form of limited collaborative effort or common interest. Irrigation communities (Crown 1987) or platform mound communities (Fish and Fish 1984) are concepts used to convey the idea of people uniting for one specific task. This task-limited concept of community reinforces the idea that social communities need not be residential, and also that aspects of community organization can transcend individual village groups. It may not be as useful, however, in understanding social relationships within a particular village site and that site's relationship to the broader social community.

My own research described here is a little different. Here, I begin with considering an archaeological site that seems to be a village. While appreciating that not all community members necessarily resided in the village itself, and that of course individuals can often identify to a greater or lesser extent with multiple communities at many different spatial, social, and analytical scales, I focus on evaluating the ways in which the archaeology of a site exhibits evidence of economic activity, social reproduction, and some degree of self-identification that Kolb and Snead (1997) found most useful in understanding village communities.

Some archaeologists extend these interests into issues of social boundaries, self-awareness, and identity, such as reviewed in Duff (2002). Evidence of identity groupings appears in the archaeological record because these associated actions leave material traces in the form of artifacts, site organization, constructions, or land use patterns (Byrd 1994). However, in the time of early villages in the Southwest, nearly a thousand years ago, evidence for identity groupings may or may not have much to do with modern concepts of ethnic groups, language groups, and their current distinctions and distributions. For example, in my study area, the Spanish military and religious authorities distinguished between Tiwa-speaking and Tompiro-speaking pueblos in seventeenth-century Salinas. However, the antiquity, significance, and continuity of these categories for people who lived many hundreds of years earlier remain open questions.

Similarly, the spatial scale of a community may not coincide with a single geographic locale (e.g., Potter and Chuipka 2007). For instance, a single village may well include more than one main identity group, especially in the context of migration. The Hopi-Tewa village of Hano (Dozier 1966) is one well-known ethnographic example; Stone (2000) and Fowles (2009) present archaeological examples. Conversely, language communities can be very large in scale, including multiple villages across a large territory (Ford et al. 1972).

Investigating Archaeological Village Communities

In practice, archaeological village sites are identified by a spatial cluster of structures. These structures may have had a variety of functions, including residential housing, storage rooms, or ramadas providing temporary shade. Spatial clustering of structures may represent seasonal reoccupation of a given locale, or a permanent village site occupied by several households. Common to all these situations, however, is the presence of a built environment, a human investment of time and labor resulting in modification of the cultural landscape.

The spatial organization of structures and settlements can be viewed as the result of dynamic cultural processes, both expressing and also reinforcing social values (Rapoport 1969, 1982, 1988). Evidence of structure size, layout, and spatial organization within a village can help us reconstruct how people classified, defined, and used space. Evidence of the built environment can also tell us about the size, composition, and organization of social groups (Schiffer and McGuire 1992; Low 2000; Fisher 2009); the nature of economic and political power (Nielsen 1995; Dovey 1999); the symbolic meaning attributed to the landscape and to aspects of society (Eco 1980; Ewart 2003), and even to the performance of everyday life (Hall 1966; Goffman 1971; Hanson and Hillier 1982).

The study of the built environment is itself a whole topic within anthropology, and various lines of research come from cultural anthropology, cultural geography, and urban planning as well as archaeology. In the studies referenced above, the built environment includes architectural features such as buildings, sidewalks, fences, patios, and plazas and their placement within a natural landscape. It also includes the spaces between these features: the passageways, open spaces, shadowed areas, vistas, and skylines. The suburban American's current interest in feng shui also picks up on ancient Chinese Taoist ideas of how architecture, interior design, and the landscape together can create a mood of harmony or disharmony.

Studies of the built environment thus attract attention from a wide variety of scholars, including interior designers, landscape architects, architects (of course), city planners, and even people who study environmental psychology. Applications of these studies range from how a pilot's cockpit might be arranged to reduce human error to studies of how the layout of the built environment can speed up or slow down pedestrian traffic and affect human behavior in shopping centers, restaurants, taverns, waiting rooms, or public parks (e.g., Sommer 1969; Goffman 1971).

Archaeologists who are interested in space syntax form another group of researchers who have similar interests at a larger scale. For example, archaeologists use observations of structure size and layout to inform our understanding of the size and composition of households or other residential units (Cameron 1999a; Cutting 2003). The size and spatial organization of rooms and room blocks within a village site, and vis-à-vis other structures and features, may argue for organizational principles that shape identities of social groups larger than the individual nuclear family unit (e.g., Liebmann and Preucel 2007) or allow us to evaluate possible evidence for social boundaries between such groups (Reed 1956; Barth 1978; Stark 1998; Wimmer 2008; Schachner 2010).

At a smaller scale, room size conveys information about the expected use by an individual or a group; other variables such as the size of the roofed area within a site provide, among other things, an estimate of total group size. Open spaces between buildings convey important information regarding encounter rates, traffic flow, and the size of the social groups utilizing these areas. At a larger spatial scale, the placement, location, and orientation of the village site within the physical and social landscape carries information about the larger cultural landscape of the village community, and its possible relationships with others.

In the cases where community members are not necessarily coresident, we can also consider whether a given individual or family is restricted to participation in only one village community—that is, the extent to which membership and/or participation is bounded, and the character (such as the permeability) of such boundaries, the difficulty of approaching or traversing these boundaries, and the relationships between individuals on either side of that social boundary. These and other factors affect whether individuals can legitimately identify with more than one group and the ease with which individuals manage their various membership identities. This problem of group boundaries occurs particularly where group sizes are small (that is, unsustainable as a biological mating network), where groups are strongly bounded or mutually exclusive, or where exogamous rules of marriage dictate or prefer marriage ties outside one local group.

The built environment provides these clues regarding how people in the past perceived, structured, classified, and used space. Of course, our interpretations are affected by current assumptions about families, households, and other groups, and their different activities and spheres of influence. Ideally, these assumptions are informed by cross-cultural ethnographic research (e.g., Kent 1990), but even so, their extension to more ancient societies may prove problematic.

This interpretive issue can be seen in the variety of concepts that archaeologists use to puzzle through the evidence. For example, researchers define several dimensions of space use. They may contrast public/private space, domestic/communal, domestic/corporate, or a number of other combinations. Often, however, these and other social concepts regarding the built environment are not necessarily unambiguously defined or consistently used. For example, within the literature of the built environment, a public space may refer to access, as for a building open to the public; it may refer to the size of the space and its ability to accommodate many people, as in a public square; more confusingly, public may also convey official or group representation, as when a political dignitary makes a speech, even when the audience is not particularly large.

Similarly, scholars often use the concept of domesticity to refer to matters concerning the household or resident group (e.g., Yanagisako 1979). Thus, one often sees the term domestic compared with public (or official) decisions, which affect a larger group. Domestic affairs are often equated with interior spaces in residential structures, as opposed to corporate affairs (or topics having to do with a defined group's activities) taking place in "public" (whether indoors or out), but in fact the distinction is often not so simple. Some scholars of the built environment more cautiously refer to "visible" or "non-visible" space, but these ideas may or may not be equated with exterior or interior space.

In fact, Bowser and Patton (2004) show that these different dimensions of space use vary independently; their ethnographic study in Ecuador demonstrated that almost every household within a village shows material evidence of interactions along a continuum ranging from intimate (private matters, concerning only the household) to social (a larger group) and public (official) interactions (Bowser and Patton 2004:176). This situation, they note, is not uncommon for societies in which political power is relatively decentralized and embedded in daily life (Moore 1996)—that is, the situation we might expect for the early Pueblo Southwest.

In this study, I will use the concept of village community to investigate the local and specific expressions of early village societies in central New Mexico. For each time period I begin with the archaeological evidence of artifacts, structures, and features—that is, a possible village site. I recognize that not every member of the village community necessarily resides in the village site at all times, and that the composition of the residential group might vary over time. That is, some villages might represent a single tightly knit and durable coresidential group; other villages may function more as a seasonal or year-round core residence or central place to a dispersed population.

This concept of a central place is not the strictly defined economic and political concept from Central Place Theory or the concept of "community centers" that researchers in the Mesa Verde area use to describe large residential sites or those with public architecture (e.g., Glowacki 2010:206). Rather, I am using the concept in the sense of a modern small town that forms a social and cultural focal point for outlying farmers or ranchers, a role that exists apart from its economic or political role vis-à-vis modern state and federal government administration. The historic town of Gran Quivira, for example, played such a role in that it formed the nexus of social interaction and identity for farmers within a much larger region that included parts of two counties (Link and Link 1999; Herrman 2003). Distant community members may participate only occasionally or peripherally with events at the village site itself, but may identify with the community for some purposes. For example, some outlying ranch families' participation in the community of historic Gran Quivira might have been limited to sharing burial space within the designated local cemetery, while other families were much more involved in the cultural institutions, such as the churches or public schools (Link and Link 1999). Still others may reside much farther away, yet retain social, cultural, political, and/or financial interests located within the town itself. In this idea, the village site as well as the village community may translate best to a concept of "hometown" or "ancestral home." A hometown is a concept of affiliation that operates through long-term social relationships and also people's history of interaction and experience with the natural as well as cultural landscape. In some respects, then, this exploration of early village society and community also considers how we might understand the varied social relationships that we expect to see within and also beyond the village site itself.

Archaeological Models of Early Village Society

Bandy and Fox's (2010) heuristic concept of early village society focuses on four factors: first, an economy with the potential for intensified food production (usually based on agriculture); second, daily life that includes some sort of relatively permanent residence in clustered, nucleated, groups; third, there is some assumption of the relative political autonomy of the village or settlement cluster; and fourth, nearness in time to the origins of sedentary village life, that is, early in the archaeological record. The latter two criteria distinguish between early village societies such as ancient Pueblo Indian society in the Southwest, where villages existed within a world of other villages, and those representing (for example) feudal, peasant, colonial, or modern villages that operate within a larger political economy.

Although different researchers emphasize numerous lines of evidence, and varied organizational concepts, I see two general research themes running throughout the discussion, regardless of the time period or geographic area. First, archaeologists are interested in the nature and operation of subgroups within an identified village site. The second important general research question considers the social relationships of these subgroups in relation to one another and to the village community as a whole. These research themes are often expressed as issues of social differentiation and group integration at different social scales. Existing archaeological models relate these concepts specifically to aspects of the built environment within a particular village site. Below, I explain some of the common research questions involved in these models.

Issues of Internal Differentiation

Assuming some basic nuclear family structure as a minimal social unit, then the issue of internal differentiation concerns the role of this unit within the social organization of the village community and the correlation between this social unit, residence, and economy—that is, the relationship between the nuclear family and the household (Wilk and Rathje 1982; Roth 2010). In archaeology, the household is usually considered to be a residential unit, representing a group of people sharing the tasks involved in production and consumption, and also child-rearing. The definition of households, assessments of this concept's utility in understanding social interactions, and utility of this concept in interpreting the archaeological record is of considerable interest for anthropologists as well as archaeologists (e.g., Wilk and Rathje 1982; Netting et al. 1984; Wilk and Netting 1984; Netting 1990; Blanton 1994). In the field of Southwestern archaeology, issues of household organization have been particularly important among scholars studying early pithouse sites and small pueblo sites (e.g., Lightfoot 1994; Hendon 1996; Bernardini 2000; Wills 2001, 2007; Roth 2010).

Archaeological applications of these ideas emphasize the size and composition of the group that people consider to be the residential unit, the economic unit, and the risk-sharing unit, and also how these groups relate to kin groups such as the nuclear family, extended family, lineage, or clan. While some Southwestern archaeologists have investigated ancient Pueblo Indian social organization and the origin of lineage-based kin groupings (e.g., Hill 1982; Bernardini 1996; Peregrine 2001), other studies focus more generally on evidence for residential corporate groups of varied size and composition. More recently, other archaeologists have added another variable to this mix—that of social distinctions between pueblo village founders and more recent immigrants (Riggs 2007). For all of these issues, archaeologists look to the built environment—particularly residential structures and storage features—to address some of these issues.

The size and composition of residential households and small-scale corporate groups is important to the study of early villages because, as Flannery (1972, 2002) has noted, in a society where food is widely shared, both risks and rewards are spread throughout the population. As a result, there is less inequality or differentiation within the group as a whole. This arrangement is common among basically egalitarian foragers and is illustrated with abundant ethnographic evidence among observed forager societies (e.g., Leacock 1983; Whitehead 1993).

The disadvantage of this system from some points of view is that "there is little incentive to intensify production in such societies, since whatever is produced must be shared" (Flannery 2002:421). The institutionalized practice of sharing, which is generally beneficial within the context of mobile foragers, can become a problem for nuclear family households within the new social environment of village life. Whatever original factors may have initiated an increase in residential sedentism, we can expect that this change will create a new social and economic environment and affect customary relationships and practices such as sharing (e.g., Testart 1982).

This issue of sharing and not sharing becomes particularly important when relatively sedentary groups become more dependent on agricultural production, because planting and harvesting by definition involves overproduction at harvest and stockpiling seed for the next planting season. This seed stock represents a new kind of resource within the basic way of life of mobile foragers—seeds are food, but this food cannot be eaten. Widespread and immediate food sharing, as practiced among mobile foragers, is thus counterproductive in a farming context. Sedentary village farming as a way of life therefore poses some profound challenges to ancient and embedded social institutions that concern not only mobility used for foraging but also basic assumptions about social relationships and their attendant obligations and expectations (see Bender 1978; Lightfoot and Feinman 1982; Schachner 2010).

In the case of early agricultural village societies, Flannery (1972, 2002), among others, points out that if food production, storage, and consumption are activities located within a specific and bounded group, such as a household, each group assumes the risks, costs, and also benefits associated with food collection or production. According to this logic, group members have an incentive to intensify production, since each group will presumably be able to control the fruits of its labors. Some archaeologists present this situation as a positive development, noting that the concept of private food supplies rewards those willing to work harder (e.g., Flannery 2002). Archaeologists sometime extend this concept to explain the origin of political complexity, as some individual personalities (often called "aggrandizers") recognize and actively manipulate this potential for household inequality, accumulating wealth, social obligations, and therefore political power (Hayden 1995; Flannery and Marcus 2012).

These issues of the origins of political and social inequality lie outside my primary research interest in egalitarian society. On the other hand, it is worth noting that locating responsibility for food production, storage, and distribution at the smallest possible scale in society—here, the nuclear family household—also has its drawbacks for individuals and the village community as a whole. This concept of a bounded nuclear family group in control of private storage creates a category of failure for those families who exercise poor judgment, misallocate household labor, are burdened with a large number of dependents, or experience illness, injury, or simply bad luck. Without an accepted mechanism for redefining group memberships or social boundaries, or reallocating resources (or people) between groups, this situation can lead to increasing economic inequality between households (Gurven et al. 2010).

A model of household economics that results in more than temporary and predictable economic inequality between different households is likely to lead to considerable intergroup stress between groups of "haves" and "have-nots." This scenario of economic and social inequality and the accompanying threat of geopolitical instability are familiar to most Americans today from historic and modern examples. Today, well-informed people frequently disagree whether social, political, and economic inequality is particularly desirable, or sustainable, or inevitable within and between modern nation-states. Anthropologically speaking, however, it is difficult to reconcile the idea of marked and persistent economic inequality with what we know of egalitarian society in general, whether organized as hunter-gatherers or small-scale farming villagers.

First, as Donham (1981) has cautioned, ethnographically observed households are not bounded units in small-scale society and cannot be understood as generic and equivalent social units, much less economically autonomous ones. A second issue is that in any society, individual households can be expected to vary in membership and organization, if only in terms of the temporal variation related to the domestic cycle in which some households include less productive workers such as small children or elders (Goody 1971). In such cases, social ties between households provide fluidity in the labor supply, allowing people to cope with these predictable variations. Within a community, even unpredictable events such as crop failure, accidents, and illness are common enough to form a background level of expected ill fortune, and in most cases are managed through these same familiar social ties (Donham 1981).

A third problem with this model of household self-reliance concerns a more general issue: that of the moral economy within the village community. While classic anthropological conceptions of reciprocity (e.g., Sahlins 1972:figure 5.1) proposed a social morality of sharing based on kinship distance and household membership, more recent studies provide a more nuanced view of the many different cross-cutting social relations of mutual obligation that mobilize the flow of goods, services, and personnel between groups, and their role in limiting inequality between and among different residence units or households (Flanagan 1989; Whitehead 1993; Fiske 1994; Nowak and Sigmund 2005; Nowak 2006).

Within the archaeological context of early village sites, therefore, many archaeologists reason that household control of food storage will be accompanied by evidence of social ties, sharing, and reciprocal relationships between nuclear family households (Bawden 1982; Byrd 1994; Graesch 2004). These relationships range from relatively ad hoc and temporary to highly formal arrangements of mutually recognized rights and responsibilities among well-defined and bounded corporate groups (Hammel 2005).

Issues of Village Integration

Intravillage social differentiation is also important because it is associated with a change in the size of the cooperative, food-sharing unit, from the nuclear family household to a larger corporate group. Whether early village societies are based on the nuclear family unit as the basic unit of operation, or on some extended family group based on descent (such as a lineage or clan), the sheer fact of village life involves regular association among individuals, and also between individuals who might legitimately represent or speak for larger social groups. To the extent that early villages were organized with relatively stable and well-defined corporate groups, then the set of relationships among these groups forms a second dimension of social interaction. If we expect that a single village society at one village site would include two or more corporate groups in residence, then we can also expect these group members might anticipate regular and repeated contact with individuals from other groups, across many years and even generations.

Corporate groups vary in membership rules: in some cases, group membership is inherited, but other groups consist of self-selected or invited members. Groups also vary in terms of the ease with which individuals can transfer membership between different groups, or participate in the activities of different groups (e.g., Kosse 1994). These village subgroups also vary in their significance for an individual's political, economic, and social decisions. We can also ask whether these corporate groups are relatively long-lived, and how much they structure people's activities over fairly long periods of time. For example, in some situations, strong subgroups are associated with a relatively high degree of uniformity of behavior and standardization of some activities within each particular group (e.g., Kohler et al. 2004). In addition, to the extent that these are relatively stable, bounded groups with a well-established membership, we might also expect individuals to identify more strongly with their co-members than with other individuals outside that group.

Many Southwestern archaeologists, looking at the ethnographic record of lineages, clans, moieties, and also sodalities among the historic and modern Pueblo Indian people, assume that similar corporate groups in antiquity would probably be fairly stable groups, with recognized members and fairly clear and relatively impermeable social boundaries (e.g., see Ware and Blinman 2000). The general expectation is that corporate groups would be important in the activities of daily life within a pueblo. Groups would emphasize internal group solidarity and differentiate themselves from other, similar groups within the pueblo. The logic here assumes that some kinship groups such as lineage- or clan-based households may entail sharing economic and political risks as well as benefits, with group members having a mutual stake to all members' economic well-being. On the other hand, these groups can also differ from one another in social variables such as size, demographics, knowledge, experience, and other attributes of social capital. According to this model, interactions between different corporate groups cannot be assumed to adhere to the same ideals or moral calculus that we might see within groups; these relationships are expected to follow a different social logic.

Some archaeological models of early village society in the northern Southwest seem to assume that this proposed corporate group organization would dominate social life. They postulate a fairly competitive environment of several different well-defined and bounded corporate groups, each pursuing group economic and social success, perhaps even at the expense of village-wide solidarity. In this kind of intravillage competition, smaller lineages, or those suffering some temporary setback, would lag behind larger and more successful lineages. The basis and expression of differentiation and competition may not always be the same for all groups or all time periods, however. In some studies, this observed differentiation is primarily socioeconomic (Schachner 2010); in other studies, this differentiation may be expressed within other political, ritual, or cultural contexts (e.g., Mills 2004b; Lekson 2009).

The issue here is that within a larger village community, development of such economically and socially independent, if not competitive, corporate groups could pose a serious challenge to the village organization as a whole. One of the known major threats to village continuity is fission. When disagreements between individuals and groups apparently cannot be solved through existing political institutions, people may avail themselves of the time-honored means of conflict resolution, which involves simplifying disagreements into a conflict that has different "sides," defining a winning and a losing side. In some cases, continued co-residence is considered impossible, and one side ends up leaving the village. One well-known example of village fission comes from the Hopi pueblo of Oraibi (Titiev 1944; Bradfield 1971; Whiteley 1998, 2008; Cameron 1999b). A sustainable village community therefore must develop mechanisms for balancing the autonomy of corporate group activities with the need to coordinate their activities within the village community as a whole.

In general, communities create and express a sense of unity through individual and group action (e.g., Rogoff 2003). At the village site, these mechanisms often include activities that involve participation by more than one corporate group. Activities that involve the entire village community create social, political, and economic links between individuals in different subgroups, and provide an organizational structure for subgroup interaction (see Fehr and Schmidt 1999; Mohtashemi and Mui 2003; Nowak 2006). Village-wide activities provide opportunities for individuals to share a meaningful experience and to create and recall a shared history.

Religious ideas and rituals are one kind of group activity that may develop particular meaning for participants as well as observers (e.g., Adams 1991; Plog and Solometo 1997; Chamberlin 2008). Southwestern archaeologists have also investigated how public events such as feasts operate in this way at different social scales and within both secular and religious contexts (e.g., Potter 2000; Mills 2004b, 2007; Phillips and Sebastian 2004).

Village-wide activities thus not only bridge the boundaries between potentially competing subgroups but also develop, enact, and communicate a sense of self-identification that is an important aspect of community (Kolb and Snead 1997). This process of developing a sense of group identity is often phrased as one of community building and is closely aligned with processes of social integration (e.g., Hyland 2005; see also Stanley and Willits 2004 for an example of community building addressed explicitly to ministerial settings).

Another facet of community building, however, involves differentiation of one's group from other, similar communities (Kolb and Snead 1997). In fact, Creed (2006:12) notes that extreme group differentiation and intergroup conflict seems to be highly effective in creating and defining community within each group. As a modern example, urban gangs or political cabals and factions show clearly that differentiation and opposition can create highly unified in-groups. In early village society, therefore, the process of community integration must achieve some sort of sustainable balance between the processes of developing a separate village-level identity, while still maintaining some degree of linkage with other, similar village-level groupings.

The general idea for small-scale archaeological communities is that village activities such as rituals provide a context for developing social relationships between different subgroups, but also contribute to social reproduction and self-identification of the village community as a whole. Integrative mechanisms usually include ideas and activities that create, express, and sustain a sense of group identity for members, and also demonstrate this identity to outsider observers. Where these processes of integration are socially important, we expect to see activities involving more people, with some degree of standardization in the time and nature of the performance.

For example, activities may take place in specified group areas, especially if the groups involved become large enough so that they cannot be hosted in any one individual's household space. In some cases, the social benefits of a large group experience may even justify the added labor, expense, and planning involved in building structures or spaces, and in reserving them for specialized use. Archaeologically, people's efforts to integrate a potentially diverse set of internal subgroups are often manifested in the material record as space of sufficient size to accommodate the appearance of many if not most community members. Integrative structures are usually assumed to be represented by an open space or plaza (Kidder 2004), by formal ritual structures such as kivas (Lipe and Hegmon 1989), or by other structures that seem to be special because of their large size, artifact content (Lightfoot and Feinman 1982), or location (Wills and Windes 1989). In this respect, plazas or areas of open ground, while not structures per se, are recognized as integrative structures in that they provide a locus for performance of shared beliefs in communal rituals (Adler and Wilshusen 1990; Adams 1991; Wilshusen 1991; Low 1995, 2000; Moore 1996; Shafer 1995; Chamberlin 2006, 2008, 2012). In fact, one could make a somewhat similar argument about the role of roads, paths, trails or other open-ended or linear features (e.g., Ingold 2007).

Activities that involve public performance may also be expressed in other aspects of material culture, including specific standardized symbols or special items, standardized craft manufacture, and standardized behavior in rituals or in everyday activities. For example, Kohler et al. (2004) finds increased emphasis on conformity among early village society, evidenced in increased similarity in ceramic design elements.

However, some strategies of social integration are less overtly or directly expressed in material remains. For example, social pressure to suppress or deny expressions of conflict can be more subtle or more overt, with conformity enforced through a range of individual actions and institutions including teasing, shaming, and gossip to more organized forms of social control such as witchcraft accusations and formal policing (e.g., Darling 1999 and articles in Nichols and Crown 2008; see also Graves and Van Keuren 2011). In some cases, these cultural expectations for unity and participation may form the assumed normal code of conduct in people's daily lives, but in other cases, they may be imposed from outsiders.

One relevant example from the Southwest comes from the time of Spanish rule and missionization, a time when resident priests expected the Indians to demonstrate unity and social integration by participating in Catholic rituals and adhering to Spanish law (Foster 1960). This example reminds us that although people may sometimes welcome the social stability and predictability associated with group membership, individuals or groups may often resist new or revised group norms and identities (see

Mills 2004b; Spielmann et al. 2006, 2008). On the other hand, overt resistance can trigger other consequences, sometimes quite severe. The locus of the Holy Office of the Spanish Inquisition in New Mexico at the Salinas pueblo of Quaraí (Sanchez 1993) is perhaps an apt reminder of some of the less attractive aspects of this cheerful-sounding concept of "social integration."

Models of Early Village Society in the Pueblo Southwest

In the case of the Pueblo Southwest, archaeologists tend to assume that Archaic period hunter-gatherers would likely have been organized in small mobile groups of nuclear family households. Ethnographic models of foragers emphasize the contingent and temporary nature of these groups and the ease with which individuals and nuclear family households transition among different groups. The size, composition, and membership of groups in any one hunter-gatherer encampment are therefore expected to be quite variable, and the camps themselves to be relatively temporary with ephemeral structures. On a regional scale, of course, mobile foragers commonly exhibit a fairly regular, if varied, pattern of movement across the landscape, and at a longer temporal scale, mobile groups exhibit many of the characteristics that we might associate with community. Gilman (1997), Nelson and Hegmon (2001), Nelson et al. (2006), and more recently Seymour (2009) provide different archaeological examples of community organization among mobile groups in the American Southwest.

As a result of this general shared understanding of how foragers and small-scale horticulturalists manage relatively high population movement across the landscape, when we see archaeological sites with well-defined and well-built residential structures, we commonly infer a change not just in mobility, indicating more time spent in one location, but also a change in the definition or nature of the boundaries of group membership.

Where small groups represent the economic and residential household, archaeologists generally expect to see relatively small structures such as pithouses; larger family units are associated with larger structures. The labor investment in even seasonal structures, however, implies that the number of expected occupants is known and is expected to remain relatively constant over the anticipated time of structure use, whether structures are used only once or repeatedly over time. While residential group membership may thus vary from site to site, archaeologists generally expect pithouse-dwelling groups to cope with additional group members by building additional pit structures of a set size, rather than expanding or decreasing the size of existing structures over time.

In the American Southwest, the earliest pithouse sites certainly record a degree of residential permanence in people's investment in constructing houses and other features such as storage pits (e.g., Roth and Stokes 2007, but see also Gilman 1997). These pithouse sites occur in a context of intensified plant collection and processing. In some parts of the Southwest, ancient farmers attempted to boost agricultural productivity by using a variety of planting strategies and also specific technologies. Some of these techniques involved little additional labor investment; others involved fairly labor-intensive constructions that were intended to harvest and direct runoff from summer rainfalls, reduce evaporation, or even supplement existing water supplies through construction of reservoirs, or in the desert Southwest, canal irrigation.

There are limits to these strategies of intensification in the semiarid climate of the Southwest, a problem affecting ancient as well as modern farmers (Stone 1991; Stone and Downum 1999). However, it seems clear that the early village sites of the American Southwest in general fall within the concept of early village society used here (Bandy and Fox 2010).

Implications for Investigating Southwestern Villages

Southwestern archaeologists do make fairly direct associations between the different aspects of the built environment and social organization, partly perhaps because historic and modern pueblo dwellers provided ready ethnographic analogies. However, the time depth of currently observed social arrangements of families, lineages, clans, and their physical/spatial expression in the built environment is not necessarily known, and direct analogies can be problematic when we are thinking about people who lived a thousand years ago. In addition, archaeologists are aware that there is considerable variation among the modern Pueblo Indians in social organization. Nevertheless, these ethnographic studies do show some regular patterning in the size and composition of social groups and the features of the built environment. As a result, archaeologists pay particular attention to issues of structure size, organization of structures into groupings or clusters, number of room groups, and their organization on a site, and they have long been concerned with the interrelationships of these archaeologically observed variables with the dynamics of village formation and organization (e.g., Dohm 1994; Kohler et al. 2004; Cameron and Duff 2008).

In the following sections I briefly review some of the important variables of the built environment that attract archaeological study in the Southwest and that are important for understanding Salinas sites. These notes include some observations that are likely painfully obvious for Southwestern archaeologists, but specialists in other geographic areas may appreciate this explicit approach.

Interpreting Rooms

Observations of individual structures or rooms at an archaeological village site typically include spatial data such as size, shape, proximity to other structures, and the distribution of rooms in relationship to the total area of artifact scatter and to features of the surrounding landscape. Archaeologists then typically infer room function based on a number of variables, including the spatial association of rooms, evidence of room access via doorways or windows, and also the presence/absence of internal features such as hearths and storage pits (Ciolek-Torrello 1985; Steadman 1996; Nash 2009).

In the Southwest, the technology and method of room construction is also a significant source of information regarding site history and the nature of its occupation, with semisubterranean pit structures interpreted quite differently from aboveground structures in general (e.g., Gilman 1987, 1995). Among surface structures, archaeologists commonly distinguish jacal structures from "real" pueblos, which are made of adobe or masonry construction. Jacal structures of wattle and daub (or post and clay) are considered to be ephemeral, but in many cases where people used stone foundations, the architectural remains can be more visible than those of a highly eroded adobe structure. In the case of adobe walls, pre-Spanish era construction used puddled adobe or handmade rounded "balls" or "elongate rolls" of adobe; standardized mold-made adobe bricks were a Spanish introduction (Judd 1916:242–244). Masonry construction entails use of shaped or unshaped stone in a variety of arrangements: with or without rubble fill, different proportions of mortar to stone, and patterns of stone alignment.

Room size is in fact one of the more obvious aspects of the built environment on archaeological sites (Cameron 1999). Where multiple rooms are apparent, scholars also consider mean and modal room size, and the range of variation or degree of standardization across the site and in relationship with other sites in the region. The order or sequence of room construction can provide evidence regarding the duration of occupation at a site and lacunae between occupation and reoccupation, as well as the degree to which the structures were planned in advance or represent site growth by accretion over some period of time.

What these observations actually mean in sociocultural terms is still highly debated. Over forty years ago, Kent Flannery (1972) set out some basic comparative observations about structure sizes in village sites in the archaeological record of the Near East and also Mesoamerica. Flannery's (1972, 2002) model of village organization noted that circular house settlements are commonly relatively impermanent, with seasonal or temporary dwellings. Modern peoples who live in villages of similar size and layout provide some idea of the important social variables that may correlate with these archaeologically observable differences (e.g., David 1971; Binford 1990; Lyons 1996; Wiessner 2002). For example, observations of the Hadza of Tanzania during the late 1950s showed that people established dry season camps with clusters of very small circular huts, each occupied by one or two individuals (Flannery 2002:418). People who were closely related tended to group their huts together within a particular locale.

The relationship between residential structure size and the number of residents as described here is based upon ethnographic research (Naroll 1962; Draper 1974; Kolb et al. 1985), but while the extrapolation from room size to group size is relatively straightforward, further inferences to cultural concepts such as household size and organization spur considerable debate. The problem is that concepts such as "household" do not correlate neatly with any other concept such as "family group" or "residential group" (Bender 1967; Wilk and Rathje 1982; Wilk and Netting 1984). These studies point out that categories such as "household" represent post hoc and relatively static constructs. They argue that a better understanding of households would focus on the social processes of production, distribution, transmission, reproduction, and co-residence, giving primary emphasis to action, behavior, and participation—an argument similar to that presented by Kolb and Snead (1997) regarding the concept of community.

Interpreting Room Blocks

In parts of the northern Southwest, early pueblo sites commonly include contiguous surface rooms as well as a pit structure and a discrete area of midden (Wilshusen et al. 2012). In some of the older literature on the

Southwest, this basic architectural association is called a Prudden Unit Pueblo (figure 4 shows a generalized version). Some room blocks are simple linear arrangements, but in larger sites, subgroups of communicating rooms (room suites) within an individual room block record yet another dimension of social variation. A single site that archaeologists interpret as a village site may include one or many such room blocks and a variable number of associated pit structures.

Considering the built environment of village sites in the Southwest thus involves identifying room block size and shape, not only in spatial dimensions but also in terms of room arrangement. Thus, archaeologists often describe linear room blocks, massed room blocks, or letter-shaped room blocks that resemble alphabetic shapes. In Salinas, for example, jacal site room blocks may be described as F-shaped or E-shaped, that is, with two or three projections. Room blocks can also be described in terms that American homeowners usually use for lumber measurements. In such cases, a two by four or two by six refers not to inches but to the number of rooms in each direction of the room block.

Like rooms, room blocks can also be described in terms of compass orientation, spatial proximity (nucleated or dispersed), and so forth. Where there is evidence of construction, archaeologists also report on any apparent sequence of construction, commonly in reference to corners that show where walls may have been bonded together with interlaced blocks or abutted against one another. One of the more important observations on many archaeological sites has to do with evidence of construction sequence, in terms of whether the room block was constructed all at once or by accretion over time. For this reason, archaeologists are particularly interested in evidence of so-called ladder construction, in which a standardized set of rooms is built rapidly with the long walls emplaced first and the interior partitions added later (e.g., Liebmann et al. 2005).

Despite the difficulties of identifying households or other social groups from architectural evidence (Wilk and Rathje 1982), Southwestern archaeologists have continued to experiment with different ways of interpreting the clustering of rooms, room suites, and room blocks that characterize pueblo sites. Given that we do know that matrilineal organization is characteristic of some modern Pueblo Indian groups, some researchers have tried to locate the time of origin of this form of kinship organization in the more distant past (e.g., Hill 1982; Peregrine 2001; Ware 2002).

Other studies focus more generally on the possible correlation between room block architecture and the organization of corporate groups at various scales. For example, in the Mesa Verde region, researchers in the

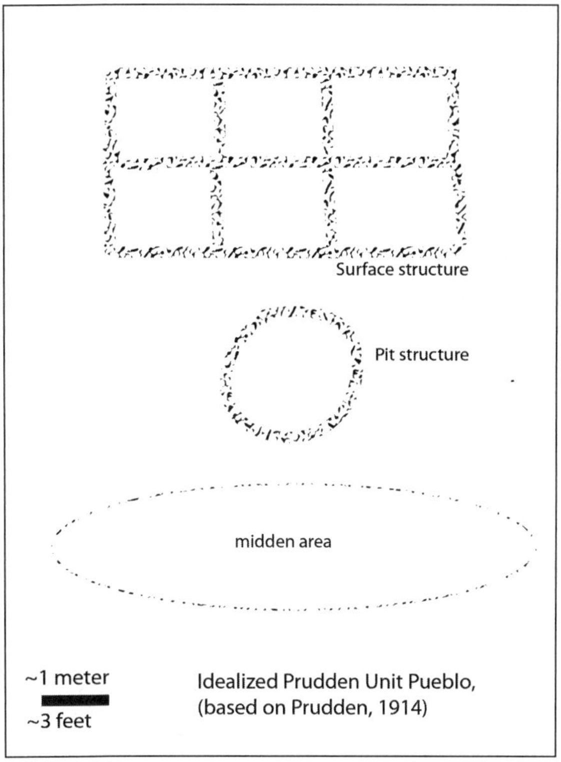

Figure 4. An example of an idealized unit pueblo from the northern Southwest (based on Prudden 1914).

Dolores Archaeological Program (DAP) began by defining dwelling units, a group of structures that includes one habitation room and one or two storage rooms. These dwelling units are associated with individual households (Wilshusen 1991). Evidence from one small Pueblo I site, the Duckfoot site, occupied from about A.D. 850 to 880, shows that a single linear room block might include more than one dwelling unit, suggesting that each room block included more than one household (Lightfoot 1994). Figure 5 shows a map of the Duckfoot site with the excavator's interpretation of room groupings based on bonding and abutting patterns observed in wall construction. In this case, although we do not necessarily have evidence of the interconnections between the different rooms themselves, the sheer fact that different room suites were constructed sequentially, in separate constructional episodes, suggests that different social groups

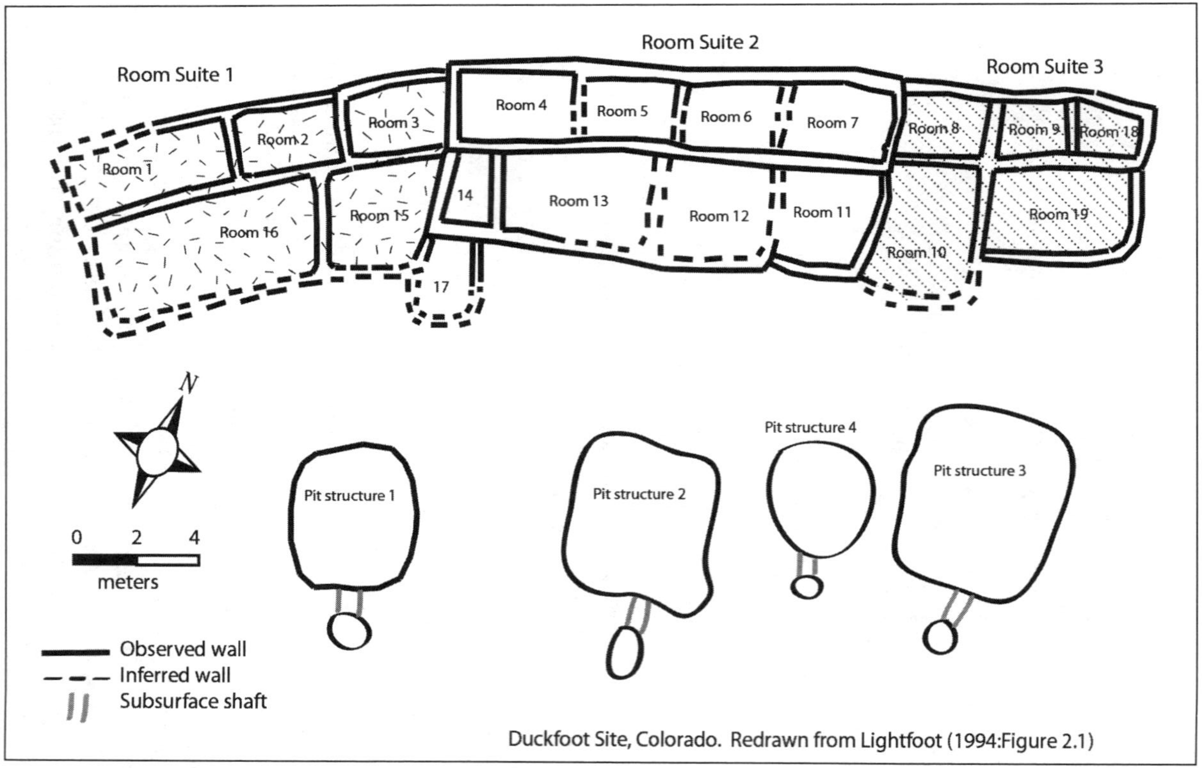

Figure 5. The Duckfoot site, Colorado, showing its incremental construction and internal organization into room suites. Redrawn from Lightfoot 1994:figure 2.1.

occupied each room suite, and were responsible for its construction. One interpretation is that the linear room blocks at this site may represent a co-operative group that consisted of multiple different households (Lightfoot 1994:160–161).

Another example of how archaeologists interpret room block architecture comes from the much larger McPhee Village, also in the Mesa Verde area and dating to the Pueblo I period. This site consists of twenty room blocks of variable length. Like the Duckfoot site, the room blocks were invariably two rooms across, but only the smaller room blocks kept the linear room block form. At McPhee Village, longer room blocks curve around, forming C-shaped or U-shaped units. In this case, Schachner (2010) used the architectural evidence of the size and constructional history of the various room blocks to investigate the development of corporate, decision-making groups. He argued that corporate group formation would be expressed in, for example, increased residence size and roofed area that could be used for storage, and increased formality of activities, including differentiation of space to separate storage, living areas, and also specialized household manufacturing areas. Other non-architectural evidence from room fill indicates that indeed the inhabitants of the larger C-shaped room blocks did engage in more activities related to manufacturing and storage. In this case, it appears that the inhabitants of McPhee Village experienced increasing social differentiation that was expressed not only in the size and shape of the larger and smaller room block groupings but also in their use of space for daily activities (see Schachner 2010 and also Wilshusen and Ortman 1999).

Other studies of room block organization do not necessarily seek such a direct correlation or connection between the observed architecture and inferred social groups or principles of social organization. Rather, these studies focus on tracing how the built environment affects more abstract variables, such as the frequency, type, and context of physical encounters among individuals and groups (Hillier and Hanson 1984), which in turn create a spatial context for social interaction (Fisher 2009). In these models, the size and interconnections among rooms within a room block (where such information is available) are important variables that allow us to identify traffic patterns, access routes, room function (residential or storage), and the internal communications that were recorded by different room groupings (e.g., Dohm 1990; Bradley 1993; Shapiro 2005).

Despite these complexities, investigation of the relationship between architecture, social interactions, and social groupings, including evidence of social differentiation, still motivates much current research in the Pueblo

Southwest, even those structures, features, and activity areas that may not be specifically or exclusively residential space.

Interpreting Site Spatial Organization

Other scholars have also noted changes in space use and architectural forms within a village site associated with changes in the size of the resident population, the degree of sedentism, and also the degree of agricultural dependence. These changes might involve structure form (pit structures or aboveground architecture), construction methods, spatial organization, or any of a variety of features of the built environment that are expressed within the area of the village site itself and its exterior spaces (e.g., Kuijt 2000; Robin and Rothschild 2002).

In all these general models, the assumption is that features of the built environment record group dynamics, including the size and composition of the group that represents the minimal social unit and whether that unit can be correlated with social groups such as the nuclear family or the cooperating extended family/lineage or is conceived more broadly as corporate groups of varying size and scale. These social changes are in turn expressed in and recorded by spatial variables of the built environment: the size of rooms and room blocks, the sequence of room and room block construction, and the spatial relationships among room blocks within a site to one another and to communally used structures or features such as the pithouse or ceremonial kiva (e.g., Liebmann et al. 2005). These basic observations of the built environment form the interpretive tool kit for investigating the internal organization of early village societies in my study area and in other geographic areas as well (e.g., Robb 2007).

Archaeology and Early Villages in the Salinas Province

This discussion of the general issues involved in studying communities, and the specific interpretations of archaeological village sites in the northern Southwest, provide the basic concepts and vocabulary needed to approach the investigation of any one particular archaeological site that we suspect might represent a village community.

These few basic descriptive variables concerning the built environment form the archeologically observable material expression of the operation of complex and versatile social processes of, for instance, aggregation, nucleation, community formation, and social integration (e.g., Cordell

1994, 1996; Lipe 1994). The variables focus on three schematic levels of variation within an archaeologically defined village site. Extensive ethnographic and archaeological research suggests a general heuristic that associates each level with a social group ranging from nuclear family to the village community.

In simplest form, we expect that small groups, probably nuclear families, are associated with small rooms or room suites, while larger corporate groups are associated with larger rooms and groupings. We expect that a village community will include not only residential structures but also storage features or structures, and other specialized areas. Some of the specialized areas may include some sort of communal facility, such as a plaza or kiva, which is suitable for activities involving different groups.

This very general model of early village society (Fox and Bandy 2010) also anticipates that there may be considerable variation in the social organization exhibited within a village community, so it should not be surprising to see a range of architectural variation in all aspects of any archaeological site that we suspect may represent a village. In this model, the relevant variation in social organization may affect the size and shape of rooms themselves, the arrangement of rooms within a single contiguous room block, including evidence of separation and interconnection and also the arrangement of room blocks within a given locale. However, to the extent that people's shared cultural expectations will affect the composition and functions of households or other residential groups, we would expect observed variation in the built environment to occur within fairly well-defined limits.

In this study, each of the following chapters presents the evidence for, and interpretations of, village society from four different archaeological sites (shown in figure 3). For each site and time period considered here, I ask two very general questions. First, given the archaeological evidence available to me, how do I know that this archaeological site with structures is really a village? Second, if indeed I am investigating a village site, what can I determine about the organization and operation of this village as a community? From this site-specific information, I then evaluate how community organization in the Salinas region may have changed over time.

My own research on village sites of the Salinas area provides an exceptional database for investigating early village society in this one geographic region and how village organization may have changed over the span of several hundred years. I also rely on the results of other people's research to supplement my excavated data with survey information and specialized analyses of ceramic manufacturing and exchange. However, a word of caution is in order. Because my excavations were originally conceived as different research projects that focused on specific, albeit related, topics, not all types of data or lines of evidence are available for each archaeological site.

Furthermore, in general, one must recall that the Salinas area is not particularly well-known. It is important to realize that for some time periods, my research reported here really does represent most of what we know about any one particular time period. This warning is not intended to discount earlier work by many other very competent archaeologists who were engaged in salvage projects as well as research studies, but will, I hope, keep the reader's expectations in line with what interpretations are in fact possible with the data available.

Accustomed to large-scale and long-term archaeological research projects in many other areas of the Southwest, some archaeologists may find it difficult to adapt to a more modest analytical scale. A more positive view would see only tremendous research opportunities available for the next generation of Southwestern archaeologists who are interested in an exceptionally gratifying set of potential research topics within an area with beautiful scenery, a deep and meaningful history, no bugs, and also a long tradition of active local community engagement in, and in support of, archaeology.

I begin with this brief review of the Paleoindian and Archaic periods, simply for the sake of orienting the discussion of early village communities within the entirety of the local archaeological record (see table 2, pp. 32–33). For each locally defined time period, I provide a basic outline of the most important studies and note where my own contributions lie. I also comment, as appropriate, on what aspects of the archaeological record proved to be particularly relevant to this study.

Paleoindian Period (about 9500 to 6000 B.C.)

The archaeological record of the Salinas region begins with Paleoindian occupation along the shores of the Pleistocene lakes in the Estancia Basin (Tainter and Levine 1987:12–26). The early Paleoindian period is actually best known from excavated Clovis period sites located much farther east on the Plains. Because of finds on these Plains sites, one often thinks of finely made bifacial spear points or fluted points as the typical Paleoindian artifact form, but in the Salinas area few such bifaces are known. Rather, choppers, bifacial knives, scrapers, and gravers dominate the artifact assemblages. Many of the stone tools of this time are made with a locally available reddish-purple quartzite; these tools are locally called the Estancia Complex (Woodbury 1993).

The best-known Paleoindian site in the area is the Lucy site in the Estancia Basin (Roosa 1956a, 1956b). Here, artifacts including a late Paleoindian Sandia point were scattered along the shoreline of a long-dried playa lake; earlier Clovis artifacts have been found on older terraces that rise above the shrinking lakebed. Other Clovis sites have been identified both south and north of Chupadera Mesa (Haynes 1955; Stuart and Gauthier 1981). Mockingbird Gap is another well-known Clovis site located south of the Estancia Basin, in the northern portion of the Jornado del Muerto. It was discovered in the late 1950s (Weber and Agogino 1968) and has been recently re-excavated (Weber 1997; Weber and Agogino 1997; Huckell et al. 2006; Holliday et al. 2007).

Paleoindian sites dating to the succeeding Folsom period have been recorded to the west and north of the Salinas region (Lyons 1969; Judge 1973), but a more recent survey found no evidence of Paleoindian occupation from Chupadera Mesa itself (Baldwin 1983; Montgomery and Bowman 1989).

The Archaic Period (about 6000 B.C. to A.D. 600)

In much of the Southwest, Archaic sites record the migrations of hunter-gatherers who moved seasonally through many different areas as they utilized a variety of plant and animal species in different environmental zones (Irwin-Williams 1979; Cordell and McBrinn 2012). The Archaic period in general represents a time of economic change throughout the Southwest, during which people began to cultivate domesticated plants such as maize for food (Cordell and McBrinn 2012). In the Salinas area, the end of the Archaic dates to about A.D. 600 (Tainter and Levine 1981:26–31).

By the end of the Archaic period, maize farming is well established as a possible subsistence strategy over much of the Southwest, and in some areas such as the Tucson Basin, the Zuni Mountains, and the Colorado Plateau, people were constructing relatively long-term or reusable shelters (Wills 1988b). However, identified Archaic sites in the Salinas area are more ephemeral; one finds only fairly small scatters of lithic tools and debitage across the western part of Chupadera Mesa (Montgomery and Bowman 1989). Baldwin (1983) notes that these lithic scatters are found both in open areas and also near rock-shelters.

There has been little recent archaeological work on possible Archaic populations in the Salinas region, however, and Baldwin (1983) notes that it is entirely possible that systematic survey of Chupadera Mesa would provide more evidence of use of this region during this time. It is also notoriously difficult to date shallow lithic scatters, much less to identify the activities that they represent (Tainter 1979; Brett and Shelley 1985; Tainter and Levine 1987:26–31). Thus, although there are Archaic sites that are interpreted as villages in other portions of the Southwest, in the Salinas area the period characterized by early villages occurs later in time, after people were already well acquainted with domesticated plants such as maize.

The Pithouse Period (about A.D. 600 to 1150)

In some areas of the Southwest, pithouse sites can be readily identified and even mapped using visible pithouse depressions on the ground surface. In the Salinas region, however, identifying pithouse occupation is more of a challenge because the underground portions of the structures are typically completely filled with windblown sand. Even so, pithouse sites are easier to identify from surface evidence than Archaic sites, because they usually manifest a markedly greater number of surface artifacts, including pottery as well as lithics.

My own excavations at Kite Pithouse Village, described in chapter 2, provide some basic information about subsistence and land use in the latter part of the Pithouse period. This information is sadly lacking in terms of data on architectural variation and site spatial organization, since my research project in 1986 focused mainly on recovering information about subsistence from the middens. However, excavations from this site also yielded a small assemblage of human skeletal remains, which I was later able to compare with that from a pueblo site to determine that there is little evidence of change in people's diet over time. Thus, excavations at Kite Pithouse Village proved to be crucial in establishing this one constant, allowing me to isolate and investigate the social significance of observed architectural change over time.

The Jacal Period (about A.D. 1100 to 1300)

Jacal is the Spanish (that is, New Mexican) word for a kind of construction elsewhere called wattle and daub. Jacal structures are visible archaeologically as lines of upright stone slabs partially emerging from the sandy ground surface. These foundation stones also sometimes appear on pueblo sites, where they may represent the remains of porches or sunshades (ramadas).

In the Salinas area, Jacal period sites include multiroom structures, with two or more contiguous rooms composing one linear room block. Nearly everything we know about the Jacal period comes from Caperton's (1981) reconnaissance survey of sites that were already known to local ranchers, and one recent systematic survey on Turkey Ridge, located southwest of Gran Quivira Pueblo (Chamberlin 2008). These were not my research projects, although Matt Chamberlin graciously permitted me to accompany him for a couple weeks of survey. In this study, I report on his published results, and place them within a larger context using other published surveys from nearby areas (Marshall and Walt 1984; Montgomery and Bowman 1989) and from my own observations (see chapter 3).

The Early Pueblo Period: Adobe Pueblos (about A.D. 1150 to soon after 1313)

In much of the northern American Southwest, the Pueblo III period (about A.D. 1150 to 1300) and the early part of Pueblo IV (about A.D. 1300 to 1425) experienced region-wide abandonment, population migration, and also in some areas, population nucleation into very large pueblos (Cordell 1995; Adler 1996a, 1996b; Kohler and Sebastian 1996; Spielmann 1996; Duff 1998; Amith 2005). In the Salinas region, this process is represented by population nucleation during the Early Pueblo period, with people moving from scattered local jacal groupings into highly nucleated pueblo villages made of adobe (Rautman 2000). These earliest pueblos in the Salinas area consist of four room blocks defining a square central plaza. My excavations at Kite Pueblo (LA 199) and at Frank's Pueblo (LA 9032) provide nearly all of the information that is known about this time period in Salinas (see chapter 4). The end date of this time period is marked by appearance of imported Rio Grande glaze wares on archaeological sites in Salinas. Because so little glaze ware pottery is found on these earliest pueblo sites, it seems likely that occupation at these sites ceased at about the time that people began manufacturing the earliest glaze ware types (see Eckert 2006 and the discussion below).

The Early Pueblo Period: Glaze A Masonry Pueblos (about A.D. 1313 to 1425)

By about A.D. 1300 a number of small compact masonry pueblos are scattered around the Salinas region. These sites are sometimes called "the

Glaze A pueblos" because early Rio Grande glaze ware types are found on the ground surface.

The first glaze ware vessels in the Southwest are red pottery vessels decorated with vitrified mineral paints. These early glaze types come from sites in east-central Arizona dated to the late 1200s; they are called St. John's Polychrome and later, Heshotauthla Polychrome. Along the Rio Grande, people modified this basic technology to produce oxidized and fully vitrified glaze painted wares (Mera 1935; Shepard 1936). The first of the eastern pueblos to make glaze ware vessels were probably those in the Galisteo Basin, just south of Santa Fe. Here, deposits of lead ore (galena) in the Cerillos Hills provided the source material for the lead-based glazes. People in this area utilized and likely controlled access to these sources of lead. These same groups produced most of the glaze bowls that are found on archaeological sites throughout much of the Rio Grande in the fifteenth and sixteenth centuries (Spielmann et al. 2006). The Galisteo Basin pueblos also set the regional standard in terms of the basic glaze formula and the overall design style of the decoration.

By the sixteenth century, people at other Rio Grande pueblos were making their own glaze vessels and the role of the Galisteo Basin pueblos declined in importance, although the basic parameters of technique and style they set continued to be widely used. During this time, people at the Salinas pueblos of Abó and Quaraí also began making their own glaze ware vessels with some local variation in glaze composition and decorative design (Huntley et al. 2007). Researchers note changes in the composition, form, and decoration of Rio Grande glaze wares by referring to an alphabetical sequence of types, from Glaze A pottery types that begin to be manufactured in A.D. 1313 through Glaze F, which was manufactured until the 1700s (Eckert 2006).

During the early Glaze A time period considered in this volume, however, the glazed vessels in the Salinas area are all imported, mostly from the Albuquerque and Galisteo basins. My own excavations at two different sites, Pueblo de la Mesa (LA 2091) and Frank's Pueblo (LA 9032), provide detailed architectural information about pueblo construction, community organization, and everyday life during this span of time (see chapters 5 and 6).

The Later Pueblos (after about A.D. 1410 to the 1670s)

By the time that Glaze C pottery is manufactured around A.D. 1430 (Eckert 2006), most of the early pueblos were abandoned. Some of them may have been reoccupied later after a period of abandonment (Montgomery and Bowman 1989), but overall the regional population nucleates into a small number of very large pueblos, each with hundreds of rooms and multiple large plazas.

The first records of Spanish contact with the Pueblo Indians of the Salinas Province date to the time of the Spanish expeditions of 1582 and 1583; in October 1598, representatives of the largest Salinas pueblos signed Don Juan de Oñate's Act of Obedience to the King of Spain (Hayes et al. 1981:4; Murphy 1993:22). Franciscan missionary priests started arriving in 1622; they chose four of the largest pueblos to become Franciscan mission sites (see table 2). The impressive ruins of the Catholic churches and *conventos* can still be seen at Abó, Quaraí, and Gran Quivira, while a smaller chapel has been found at Pueblo Blanco (Tabirá).

The late-period occupations at these large pueblos both before and after Spanish contact do not really fit into the concept of early village society as I use it here. Accordingly, my research regarding early villages begins with some background information about the earliest sites with structures, the pithouse villages, and ends with discussion of the early Glaze A pueblos.

CHAPTER TWO

Pithouse Village Sites and Communities

Pithouse village sites record some of the earliest structures throughout the Southwest (Gilman 1987, 1997; Rocek 1995; Diehl and LeBlanc 2001; Cordell and McBrinn 2012), but the term pithouse encompasses a truly amazing variety of structural forms. Pithouses vary in size, shape, depth below the ground surface, and construction methods (Bullard 1962). Some are accessible only through a hole in the roof that doubles as a chimney; in other areas, entry to the pithouse is through a side room. In later times, pit structures within a given region became more standardized; one can generalize broadly that, for example, pithouses in the northern Southwest tend to be more rounded, while those from the Mogollon area to the south tend to be relatively square (Peckham 1976).

Pithouse sites throughout the Southwest are also quite variable in size. In the northern Southwest, residential pithouses tend to be about four meters in diameter, and may be furnished with internal features such as a fire pit, ventilator openings, and slab or adobe deflectors. There is also regional variation in pit structure size and use. Smaller residential pit structures can be about two meters in diameter; in the northern Southwest, communal pit structures such as great kivas can be ten to twelve meters in diameter (Cordell and McBrinn 2012).

Archaeological sites with multiple pit structures represent the first likely village sites in the Salinas region, as they do in many parts of the Southwest. In fact, however, we really do not know whether the people who lived on these sites operated as a community in any social or anthropological sense. In this chapter, I outline the small amount of information that we do know about the Pithouse period in the Salinas area in general, and then discuss how my own excavations at Kite Pithouse Village contributed to this understanding. I then consider what the available information about the Pithouse period reveals about village organization and social strategies that people used during this time.

The Pithouse Period in Salinas

No systematic survey has investigated the Pithouse period occupation of the Salinas region, but Caperton's reconnaissance survey records those sites that were known to local ranchers (Caperton 1981:figure 1). These include sites on the eastern slopes of Chupadera and Jumanas Mesas. Three other pithouse sites are located on sandy sediments at the southern end of Chupadera Mesa (Caperton 1981). Stuart Baldwin (1983) reports at least one pithouse site in the northern Tompiro Division near Abó Pueblo, but in general, we know very little about Pithouse period sites and settlement pattern. Figure 6 shows the location of these known Pithouse period sites.

Based on the ceramics present, these sites date from about A.D. 600 to 1150. We really do not have good information about potential pithouse occupations in other areas of the Salinas Province, but from the limited excavation data available, we can distinguish an early period and a late period of pithouse use in the area near Gran Quivira. The early period begins around A.D. 600 and lasts until A.D. 900 or so. This early period is distinguished by deep circular pit structures, about four meters in diameter, with internal features. The later period dates to after A.D. 900 and lasts until about A.D. 1150. Structures at the later sites tend to be squarer in shape and smaller in size, only about two meters in diameter. In both time periods we find thick midden deposits inside and also outside structures. From general studies of midden accumulation, we know that these thick deposits indicate that people were using and reusing specific favored locales over some period of time (e.g., Varien and Ortman 2005; Beck 2006).

Identifying Pithouse Sites in Salinas

In some areas of the Southwest, pit structures and pithouse sites can be readily identified during survey by visible surface depressions. In the Salinas area, however, sand washing down from the mesa uplands and blowing across the Medanos Plains has filled most pithouse depressions over time,

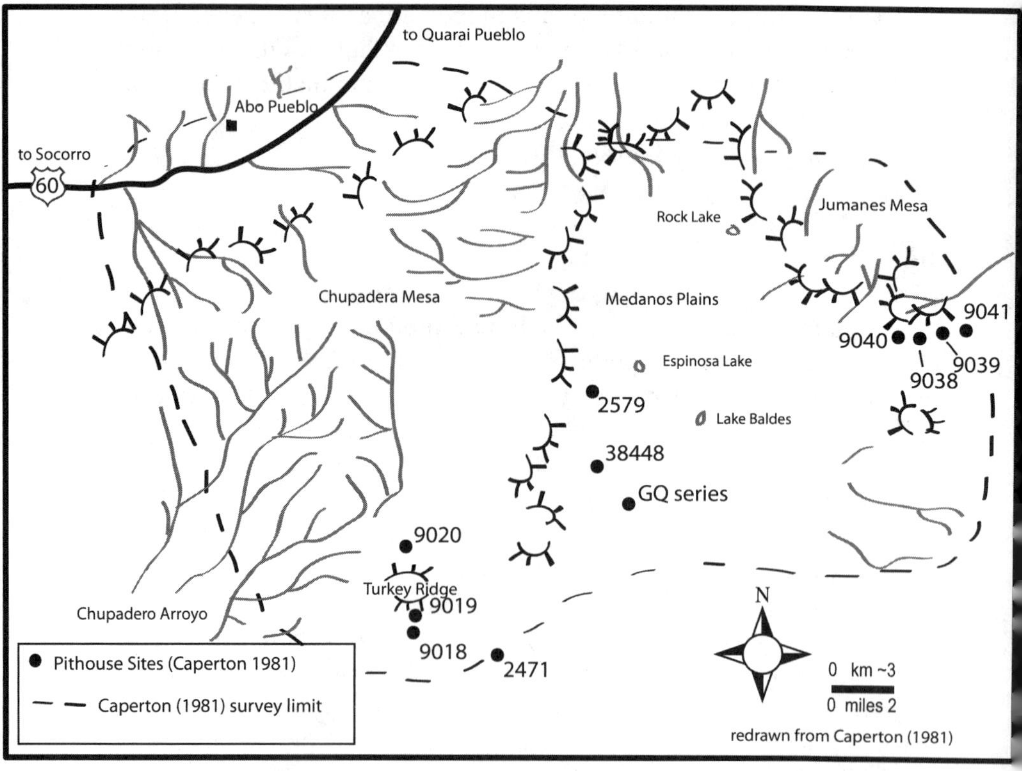

Figure 6. Location of pithouse sites in the Salinas Province. Modified from the base map shown in Caperton 1981:xii.

obscuring any trace of the structures on the ground surface. As a result, the surface evidence is limited to scatters of lithic and ceramic artifacts, and it is difficult to estimate site size or the number of structures based only on surface evidence. More detailed information about the known pithouse sites discussed here comes from the accidental exposure of buried pit structures within arroyos or from construction disturbance.

Earlier researchers also sometimes recorded "pit features" on or near later sites. Many of these pit features include small groups of multiple shallow basins near pueblo sites. These pit features are probably best interpreted as water retention features or reservoirs (see Toulouse 1945). Pit features located within the central plaza are also found on several of the early pueblo sites (Caperton 1981). Some archaeologists, including myself, have assumed that these pit features within plazas were ceremonial rooms or kivas. While some of these pit features at the later pueblo sites are indeed confirmed kivas, this interpretation cannot be assumed for the time of early villages considered here. Based on my own excavations, I suspect that most likely they represent water storage features or cisterns. Use of small reservoirs and also deeper cisterns among the Salinas Indians is known from Spanish records at the large late pueblos (Toulouse 1945; Howard 1959, 1981), and excavated evidence suggests that this practice had considerable antiquity. Chapter 4 discusses the evidence for cisterns at an early pueblo site called Kite Pueblo, and in chapter 5, I discuss the interpretation of two plaza pit features at a later site, Pueblo de la Mesa.

Pithouse period sites thus rarely can be identified or mapped using surface evidence of recorded pit features (Beckett 1981:74). Instead, they are most commonly recognized simply by the concentration of brown ware and also white ware pottery on the ground surface. Brown ware ceramic types are associated with the earliest village sites in the highland Mogollon region to the south and southwest, and can date as early as A.D. 300 on some sites (Kelley 1984). Types such as Jornada Brown ware, El Paso Brown, or Alma Plain are found in the Salinas area and are generally assumed to be utility ware or cooking pots. Later brown ware pottery (after about A.D. 800) shows the use of decorative external surface corrugation. In the Salinas area, a type called Corona Corrugated represents the bulk of the utility pottery on all but the earliest sites. In general, the relative proportion of the vessel surface that exhibits corrugations increased over time. Earlier vessels may show this banding around the neck, but later vessels are made with corrugations covering the entire vessel neck and body. Later vessels also are generally larger in size and more globular in shape (LeBlanc 1982:117–119).

In the southern portions of New Mexico, painted brown ware becomes common after about A.D. 1000, but in the Jornada (northern) Mogollon, and in the Salinas region, this painted brown ware is unknown. Instead, at this time some of the brown ware vessels in northern Mogollon and also Salinas sites exhibit a new kind of interior finish—a black, highly burnished finish called smudging (LeBlanc 1982:118). The purpose for smudging is debated, but Rogers (1980) has argued that it may help reduce permeability (that is, help in water retention) and may also inhibit fungal or bacterial growth inside the jars (Rogers 1980; Mueller 1995).

Some Salinas corrugated jars do show this interior smudging. More commonly, however, corrugated jars are simply blackened on the inside and also on the outside from heat and soot. Many show multiple visible layers of soot on the vessel exterior, suggesting long periods of exposure to cooking fires. Brown ware vessel fragments on Salinas sites usually come from wide-mouth jars. Some jars represent a wide-mouth rounded cylinder shape that would hold about four liters and fit into our modern expectations for a cooking vessel. Others are quite large and globular, nearly spherical in shape, with a narrow collared orifice. These large ones can hold twenty to forty liters of liquid (Hayes et al. 1981).

The most common painted pottery at pithouse sites in the Salinas region is a type of white ware pottery with black painted designs. White or gray ware pottery is generally considered to be northern (that is, northern Southwest) in origin, but was manufactured over a large region, with local variations in clay, paste, and paint composition.

Early painted pottery types found in Salinas pithouse sites include Red Mesa Black-on-White, San Marcial Black-on-White, and Chupadero Black-on-White. Red Mesa Black-on-White was manufactured from about A.D. 850 to 1125 (Wiseman 1986:5); the dates for San Marcial Black-on-White are about A.D. 750 to 950 (Marshall and Walt 1984:37). Chupadero Black-on-White appears somewhat later, with manufacture beginning around A.D. 1100 (Wiseman 1986:4; see discussion in Clark 2006:67–68). The presence of brown ware and also painted black-on-white sherds are thus commonly the most visible surface indicator of a pithouse occupation in Salinas.

The Known Pithouse Sites in Salinas

A limited amount of excavation at three different pithouse sites provides the information considered in this chapter. These three sites are all situated on the slopes on the eastern side of Chupadera Mesa (table 3). In the order of their investigation they include: (1) excavation at a site that I have termed the Pipeline site (Green 1955; Fenenga 1956); (2) surface survey and some limited testing that identified pit structures within the National Park Service property boundaries at Gran Quivira (Beckett 1981); and (3) my own excavations at Kite Pithouse Village (Rautman 1990, 1991, 1993b). Since the pithouses at Gran Quivira are located less than 2 km away from Kite Pithouse Village, it is conceivable that the whole lowland area around Gran Quivira/Kite ranch represents a very dispersed settlement (suggested by Beckett 1981:91)—an idea that would require much further research.

The pithouse sites considered here are all located in open territory on the Medanos Plains. In this area the loose surface sand is variable in thickness, ranging from about 50 cm to more than a meter in depth, and is underlain by a less permeable deposit of white caliche and/or fine-grained

Table 3. Pithouse period sites in the Salinas Province

Site	*Description*	*Reference*
Northeastern Region Near Jumanes Mesa		
LA 9038 LA 9039 LA 9040 LA 9041	Location: by Jumanes Mesa Ceramics: These sites contained Jornada Brown ware, Puerco Black-on-White, Socorro Black-on-White, Chupadero Black-on-White, Casa Colorado Black-on-White, Red Mesa Black-on-White, San Marcial Black-on-White, Kaithuthlana Black-on-White, Mimbres Black-on-White, Wingate Black-on-Red, Lino Gray (one sherd), Pinnawa (one sherd), and Corona Plain (2 sherds)	Caperton 1981
Southern Region Near Chupadera Mesa		
LA 2471	Location: 13 km southwest of Gran Quivira	Bradfield 1929
LA 2579 (Pipeline site)	Location: just northwest of Gran Quivira Structures: 2 pit structures, one 4.1 m in diameter, 1.5 m deep; the second measuring 4.6 m in diameter, 0.8 m deep Approximate floor area: 12–14 sq. m Interior features: hearths, postholes Exterior features: 5 bell-shaped storage pits, each 1–2 m in diameter Ceramics: Jornada Brown ware, Alma Plain, Lino Gray ware, Kana-a Black-on-White Estimated date: A.D. 600 to 900 (early pithouse period)	Green 1955; Fenenga 1956; Vivian 1979
LA 9018 LA 9019 LA 9020	Location: the southern region of Chupadero Mesa Ceramics: These three sites contained Jornada Brown ware, Puerco Black-on-White, Socorro Black-on-White, Chupadero Black-on-White, Casa Colorado Black-on-White, minor amounts of Mimbres Bold Face, and San Andres Red-on-Terracotta	Caperton 1981
GQ 7	Location: west-northwest of San Buenaventura mission church Structures: one pit house or an early surface room	Ice 1968; Beckett 1981
GQ 8	Equivalent to GRQU 2 Location: on National Park Service property Structures: 2 pit houses and a surface structure Pit structure dimensions: 2.2 m x 2.5 m, or 5.5 sq. m Interior features: central hearth; four posts in the corners	Ice 1968; Beckett 1981

Table 3. (continued)

Site	*Description*	*Reference*
GQ 9	Location: on National Park Service property, northwest of "the main mound" (that is, Mound 7)	Ice 1968; Beckett 1981
GQ 14	Location: on National Park Service property, south of the (first) visitor center	Beckett 1981
GQ 21	Location: on National Park Service property, south of the major mound Structures: possible small pithouse Ceramics: "early sherds"	Beckett 1981
GQ 23	Equivalent to GRQU 14 Location: on National Park Service property Structures: 2 pit houses and possibly a surface structure Ceramics: Chupadero Black-on-White, Corona Corrugated, glazed body jar sherds Pit structure dimensions: 2.1 m x 2.8 m, or 5.9 sq. m Interior features: central hearth; four posts in the corners	Ice 1968; Beckett 1981
LA 38448 Kite Pithouse Village	Location: northwest of Gran Quivira Unit Structures: several small pit structures, adobe-walled surface structures Occupation: ca. A.D. 900 to 1200 (later pithouse period) Dimensions of Structure 1: ca. 4 m diameter (12 sq. m; interior hearth and postholes) Dimensions of Structure 2: ca. 2.5 m diameter (about 4–5 sq. m)	Rautman 1990, 1993b; Brouwer and Rautman 2013

red silty clay. In this kind of setting, surface water percolates fairly rapidly through the loose sand, but is retained for longer periods of time atop the less permeable stratum below, where it is available for plant use. Even today, the darker green grass on ranchland indicates broad areas where their roots can tap this subsurface moisture. Scattered playa lakes across the Medanos Plains mark local topographic low areas of the underlying caliche; in some places local ranchers have created shallow retaining ponds in these locales to take advantage of the natural subsurface drainage. It seems likely that the Pithouse period farmers were aware of this phenomenon, and situated their settlements at the base of Chupadera Mesa where soil moisture was retained and where rainfall was augmented by surface runoff from the mesa.

The first site to be professionally investigated was the Pipeline site (LA 2579, Fenenga 1956). It is located near the base of the east slope of the mesa, about three kilometers from Kite Pithouse Village and within sight of Gran Quivira Pueblo and the Spanish mission church. In 1953, the Permian–San Juan Pipeline project began construction across Chupadera Mesa, and workers exposed some features and pit structures along a linear trench (figure 7). The archaeologists associated with the project did some salvage excavation of five storage pits and one residential pit structure, listed in table 4. From the excavated information about pit structure size, and the appearance of the surface topography, they estimated a total of nine pit structures at the site (Fenenga 1956).

The next year, archaeologists from Texas Tech University excavated a second residential pit structure (Green 1955). The two excavated pithouses each measured over four meters across and about a meter deep; both had internal features such as central hearths and roof support postholes (figure 8). A sketch of how this kind of pithouse might have looked when it was roofed with upright posts and a wood and earthen superstructure is shown in figure 9. A mass of charred maize from near the floor of one of the pit structures (Green 1955) shows that the people at this site were using, and likely growing, domesticated crops such as corn. The large pit storage features suggest that people relied extensively on stored foods for at least some portion of their occupation.

Ceramics at the Pipeline site included Mogollon pottery types such as Jornada Brown ware and Alma Plain, and also Anasazi (northern) types such as Lino Gray and Kana-a Gray wares. These gray wares and the presence of early white ware types such as San Marcial Black-on-White led the investigators to conclude that the site was likely contemporaneous with Basketmaker III period populations in the Rio Grande area, beginning about A.D. 600 or 700, and continuing to A.D. 900 at the latest (Green 1955; Vivian 1979:143). For nearly two decades, this site represented the total information available about the Pithouse period in the Salinas region.

The second professional study of pithouse sites in the region included a field inventory and a summary of unpublished data regarding survey and testing of sites within the borders of the Gran Quivira unit of Salinas Pueblo Missions National Monument (Beckett 1981). The pithouses identified in this study were quite different from those at the Pipeline site. The Gran Quivira pithouses were much smaller, and some were apparently contemporaneous with nearby surface structures made from puddled adobe. Two archaeologists from the National Park Service, Ronald Ice and Douglas Scovill, had argued that these small pithouses were later in time than the Pipeline site pit structures. Based on the ceramic types

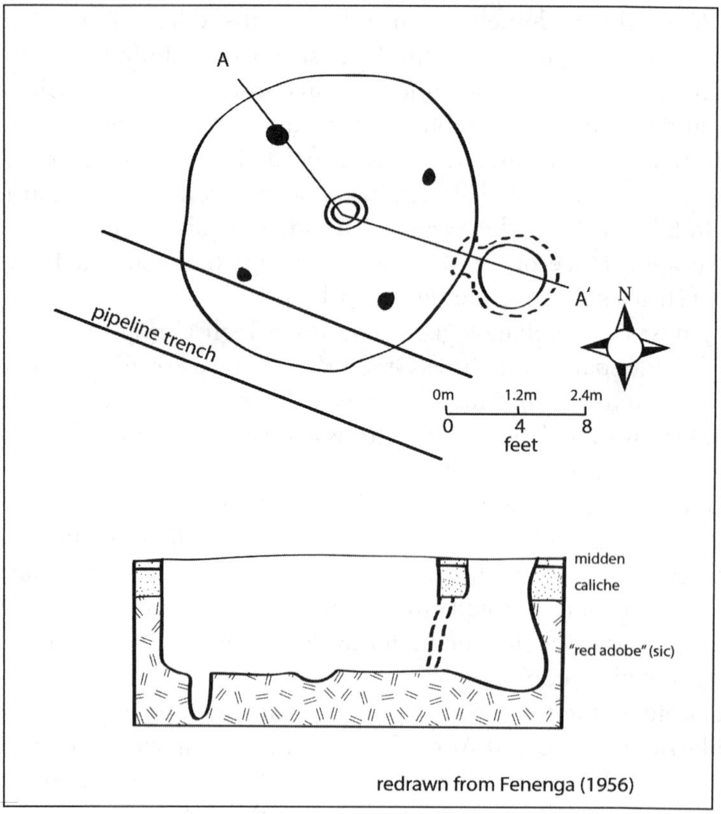

Figure 7. Plan and cross-section of a pithouse at LA 2579. Redrawn from Fenenga 1956. The "red adobe" identified in Fenenga's (1956) drawing is better described as culturally sterile silty fine sand, with calcium carbonate (caliche) deposits near the ground surface.

present, they dated these structures to after A.D. 900, or A.D. 1200 at the latest (Ice 1968; Beckett 1981).

Another pithouse structure near Gran Quivira was similarly late in date, but was stratigraphically below some adobe wall stubs, indicating that it predated the construction of the surface structure. This information makes it clear that some pit structures precede surface structure, but that there is also some overlap in their time of use (Beckett 1981).

The small pit structures identified near Gran Quivira were square in shape, with rounded corners, similar to those in the Mogollon highlands to the south. They did have interior features such as a central hearth, a

Table 4. Features at LA 2579 (Pipeline site)

Feature	*Location*	*Interpretation*	*Size (meters)*	*Fill*
1	Easternmost feature	Basin-shaped fire pit, dug into midden, intrusive into red clay	.48 m across, .3 m BGS*	Ash and small bits of charcoal, no artifacts
2	12.5 meters west of Feature 1	Pit dug into midden and into caliche below	1.9 m deep, 2.1 m across	Mixed midden and sand, 8 sherds
3	21.6 meters west of Feature 1	Bell-shaped storage pit, dug into midden and red clay	1.8 m across, burned walls	Mixed midden soil, no artifacts
4	25.9 meters west of Feature 1	Two storage pits, dug into midden, caliche, and also red clay	1.5 m deep, 1.5 m diameter; the bell pit is 1.3 m deep	Mixed midden soil; 3 sherds
5	126.8 meters west of Feature 1	Pithouse		Midden with 283 sherds and artifacts on the floor
6	26.5 meters west of Feature 1	Bell-shaped pit	.91 m deep, .91 m in diameter	Midden, 2 sherds, 2 lithics

*BGS is below ground surface. The measurements were originally presented in the English system (feet) in Fenenga (1956) and Green (1955).

ventilator shaft, and interior postholes that likely held the wooden beams to support the roof (Beckett 1981). In the smaller pit structures, however, the postholes were located quite close to the structure's edge, which would sensibly enough maximize the open space of the interior. Although these small pit structures may seem quite cramped, in fact similar small pithouses are known from other sites along the Rio Grande valley, near the modern town of Belen (Ferdon and Reed 1950; Weber 1973).

These two studies established that Pithouse period groups were likely at least seasonal residents of the area and that they were likely farming maize and relying on stored foods for part of their time of occupation of the site. In addition, the fact that some but not all of the features and structures were filled with domestic debris indicates a degree of patterned continuity over time, with repeated occupation concentrated in selected locales.

These two studies also established that there were at least two periods of pithouse occupation in the Salinas region (Ice 1968). The first one,

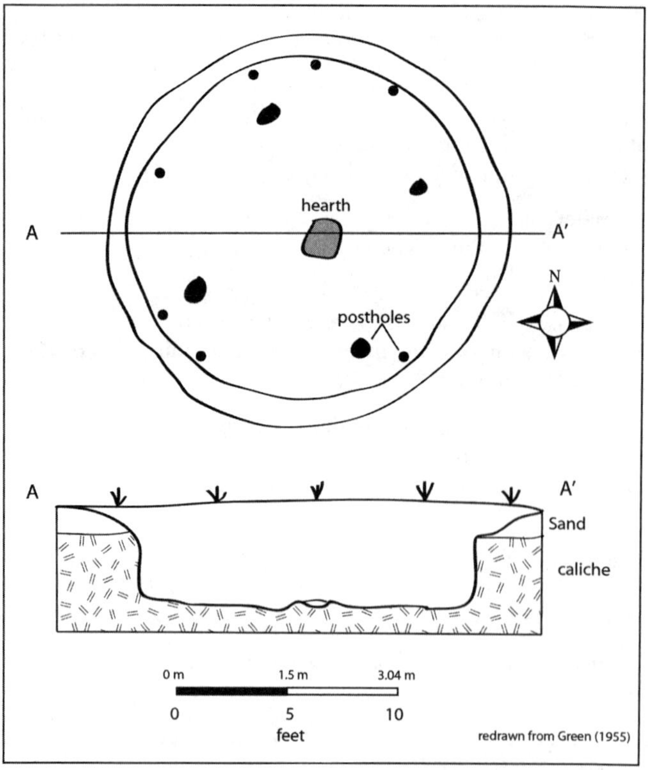

Figure 8. Plan and cross-section of a second pithouse at LA 2579. Redrawn from Green 1955. Green (1955) uses the term "caliche" in this drawing, while Fenenga (1956) uses the term "red adobe." Based on my own observations at Kite Pithouse Village, I would describe the sediment as culturally sterile silty fine sand, with calcium carbonate (caliche) deposits near the ground surface.

represented by the Pipeline site, was defined by the presence of larger, circular pit structures, with ceramics suggesting a date of A.D. 600 at the earliest. The second pithouse occupation was identified by smaller, more rectangular structures, and ceramic types that dated after A.D. 900. The single most abundant of these ceramic types on the late pithouse sites was a painted type called Chupadero Black-on-White; its production begins about A.D. 1100 (Clark 2006). Some of these later pit structures may have predated any adobe surface architecture, but on some sites, the smaller pit structures and the surface structures were possibly contemporaneous.

Figure 9. Reconstruction of an early Salinas pithouse. This particular pithouse was located in the Estancia Basin and dates to about A.D. 600. Redrawn from Murphy 1993:7.

These two studies set the context for my own excavations in 1986 at Kite Pithouse Village, which I consider in some detail here.

Excavations at Kite Pithouse Village

Kite Pithouse Village (LA 38448) was discovered in the 1980s on the ranch owned by Jack and Louise Kite. The Kite family, like many local ranching families, had a close connection with the archaeological site of Gran Quivira, which is located on a prominent limestone ridge less than a kilometer from their home. In fact, Louise Turner Kite's parents homesteaded in the historic town of Gran Quivira at the base of the hill, and she grew up in that community. The Kite ranch encompasses the historic ruins of Gran Quivira (the homesteading settlement), which had a population of about six hundred in the early twentieth century, during the height of pinto bean farming in the area (Link and Link 1999; Herrman 2003).

In this case, arroyo cutting had exposed a buried pit structure, and the Kite family recognized its importance and asked archaeologists from the Museum of New Mexico to assess the site. Patrick Beckett and Regge

Wiseman did some salvage excavation in 1982, recovering artifacts from the pithouse fill and recording the details of its construction (Beckett and Wisman 1982; Wiseman 1986). The Kite family also permitted me to test the site further for my dissertation fieldwork in 1986.

My research at Kite Pithouse Village was designed to obtain basic economic information, including an evaluation of people's use of farming as well as hunting. Because it is common knowledge that farming is a rather uncertain enterprise in the semiarid Southwest, I wanted to investigate how people coped with a high degree of environmental and climatic variability. I was not particularly interested in spatial relationships or architecture, or in village organization per se. As a result of this research design, most of my excavation focused on recovery of artifact and subsistence remains, more specifically plant and animal remains, from thick midden deposits that we identified at the site. We excavated only one small pithouse with its accompanying features, but we also found that there were larger pit structures stratigraphically underneath an adobe-walled surface structure.

While these excavations do not contribute much to the current study of intravillage social dynamics, this excavation information established some baseline data regarding early farmers in this area. These data include radiocarbon dates of the midden deposits, a sample of artifacts, and most importantly, a small burial population* that contributed dietary information confirming that the site's inhabitants were farming and eating quantities of maize (Katzenberg and Kelley 1991; Colyer 1996). In the rest of this chapter, I report on some of the particular findings that relate to village organization and combine this information with that from the other known sites to get a better idea of some of the general features of social organization during the Pithouse period.

Excavations at Kite Pithouse Village showed that this particular locale at the base of Chupadera Mesa was a locus for fairly intensive human occupation for many generations. The architecture included at least two

---

*The excavation and analysis of human burials can be a contentious subject. People around the world have very different ideas regarding death and burial, dignity and the sanctity of the body itself, appropriate burial treatments (e.g., burial, cremation, secondary burials, and the like), and the relationship of the dead to the living. Even an introductory discussion of these issues would go beyond the scope of this volume; for an example of this anthropological and archaeological discussion, see Jones and Harris (1998). Coltrain et al. (2007) provide a recent example of how human bone chemistry can help us understand ancient diet and economic organization. In the case of the burials at Kite Pithouse Village, the site was located on private land and some skeletal material had been exposed by arroyo erosion. The landowners preferred that we remove and curate any human remains that we encountered.

large buried pit structures and also an unknown number of pithouses of smaller size. A multiroom surface structure with adobe walls was situated stratigraphically above an early pit structure; its temporal relationship to the smaller pit structure is not known. The nearby arroyo also exposed evidence of other adobe surface rooms in the immediate vicinity, located just under the present ground surface and filled with windblown sand. A short distance to the west of the excavated site area, a large bell-shaped storage pit, lined with fine clay, was also exposed in the arroyo. Table 5 lists the excavated areas at the site, shown on the map in figure 10.

The midden at Kite Pithouse Village yielded abundant cultural material—typical domestic debris including ceramic sherds, lithic tools and waste flakes, animal bone, charcoal, and charred plant remains. This midden was nearly two meters deep, and our testing with a bucket auger showed that it was also laterally extensive, occupying at least thirty square meters in area (four meters by eight meters minimum). Lenses of yellowish brown sand interspersed throughout the profile recorded periods when windblown sand accumulated in low spots. These sand lenses represent times when the midden was not being used for discard, at least in that particular area. We cannot say, however, whether the whole site was abandoned periodically, or whether people were dumping their household debris somewhere else on the site for a while. However, the sheer volume of extramural midden sediment at this site indicates that the resident population was large enough, and lived there long enough, to accumulate an impressive quantity of domestic debris.

Radiocarbon analysis of individual maize kernels yielded dates that place the midden accumulation to the latter Pithouse period, after about A.D. 900 (see table 6).

Evidence of Daily Life

Artifacts as well as plant and animal remains from the excavations reveal a view of everyday life during the Pithouse period that includes a subsistence economy based on maize agriculture as well as the use of locally available wild plants and animals (Rautman 1990; see table 7).

The most ubiquitous plant recovered from the midden was domesticated maize, present in 74 percent of the samples tested. Other plant remains included small charred seeds and other fragments of locally available wild plants such as cactus and various wild berries. The most ubiquitous wild plant taxa recovered included weedy annuals such as *Chenopodium* and *Amaranthaceae* and also sunflower (*Helianthus*).

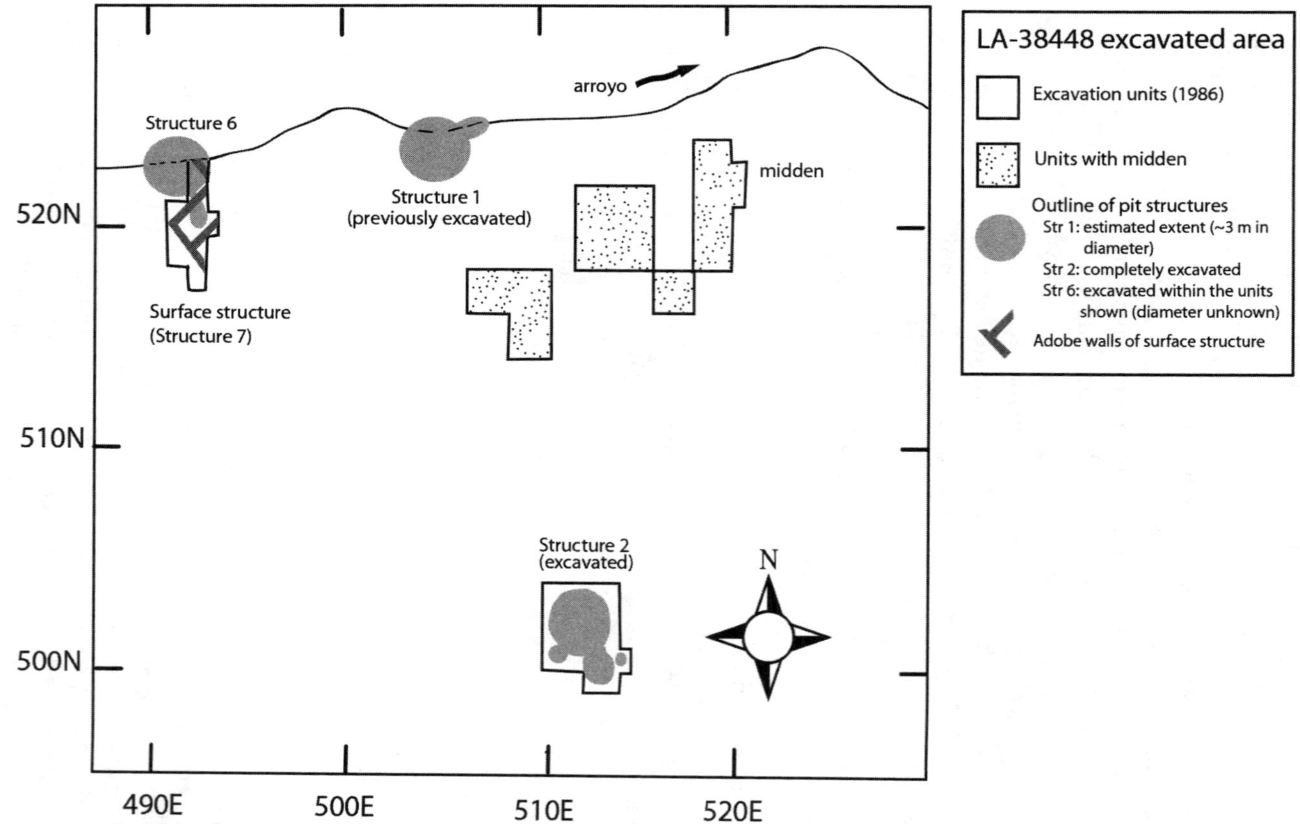

Figure 10. Excavated areas at Kite Pithouse Village, LA 38448.

Table 5. Excavated areas of Kite Pithouse Village, LA 38448

Area	*Provenience*	*Cultural Interpretation*	*Approximate Sediment Volume (cubic meters)*
A	Wiseman and Beckett's excavation (Beckett and Wiseman 1982; Wiseman 1986)	Structure 1, pithouse	Not excavated in 1986
B	510–513E 499–503N	Structure 2, pithouse	13.47
C	491–493E 517–522N	Structure 6, pithouse; Structure 7, surface structure	9.70
D	512–515E 518–521N	Midden	21.26
D	506–509E 514–517N	Midden	5.95
D	517–520E 520–522N	Midden	8.57
Total excavated sediment (cubic meters):			58.97

Faunal remains at the site indicated that people were likely eating small animals such as pocket gophers (*Geomys*), woodrats (*Neotoma*), and rabbits (*Sylvilagus audobonii* and also *Lepus californicus*), as well as larger animals such as mule deer (*Odocoileus*) and pronghorn (*Antilocapra americana*) (Elsinger and Semken 1990). All of these smaller animals are readily found in the immediate site area even today, but the fact that their sometimes charred bones were found in the midden suggested that they were intentionally discarded as food waste. Both mule deer and pronghorn are also common in the area today, although mule deer tend to range in the more wooded areas at higher elevations on the mesa itself and in the Manzanos Mountains, while pronghorn tend to graze on the lower-elevation grasslands. No non-local fauna such as fish were identified. In other words, it is unlikely that the site's inhabitants would have had to range more than a couple of kilometers in any direction to obtain meat, hides, or animal bone for tools.

Analysis of the deer and pronghorn remains, and the relative abundance of their different body parts, showed that there was little evidence that these animals were brought in from any great distance away (Lyman 1985; Szuter and Bayham 1989; Lupo 2006; Faith and Gordon 2007). However, the deer and pronghorn remains lacked many elements of the axial skeleton, that is, the vertebrae and the ribs. These body parts generally

Table 6. Radiocarbon dates, Kite Pithouse Village, LA 38448

Beta number	*Context*	*Uncalibrated age 14C (±1σ)*[3]	*Calibrated date range (A.D. ±1σ)*	*Intercept (years A.D.)*
Beta-21137	511E 502N Level 9; Structure 2, Maize cob	1090±90 B.P.	A.D. 782 to 1019	A.D. 979
Beta-21133	519E 522N Level 6; Midden, Maize kernel	790±90 B.P.	A.D. 782 to 1019	A.D. 1257
Beta-21134	519E 522N Level 8; Midden, Maize kernel	970±90 B.P.	A.D. 1160 to 1281	A.D. 1025
Beta-21135	519E 522N Level 10; Midden, Maize cob	1020±90 B.P.	A.D. 984 to1186	A.D. 999

Radiocarbon dates were determined using the AMS technique by Beta Analytic, Inc., Coral Gables, FL. Uncalibrated dates by Beta Analytic, Inc. are reported as radiocarbon years before A.D. 1950, using a half-life of 5,568 years. No corrections were made for isotopic fractionation. Calibrated dates are calculated according to Stuiver and Becker (1986). Dates were originally reported in Rautman (1990).

have little muscle mass or marrow, and are commonly the first parts of the body to be discarded when hunters butcher an animal. The lack of axial elements at the site thus suggests that at least some initial processing of animals, and discard of some body parts, occurred off-site. The location was far enough away so that the hunters did not carry the entire carcass back to their habitation site to butcher; rather, they likely field dressed and disarticulated the animal somewhere else. They discarded some parts of the body there, bringing back only selected portions to their homes, where the bones were likely further processed, the meat removed, and the bones ultimately discarded.

This kind of analysis is less useful in interpreting how hunters dealt with smaller animals, because small animals are easier to carry to the village site as whole carcasses. At Kite Pithouse Village, most of the small animal bone came from cottontail rabbits and pocket gophers. Pocket gophers are small burrowing rodents, between a hamster and a guinea pig in size. Pocket gophers as well as cottontails prefer a habitat with softer sediments, a variety of plant foods, and a relatively wooded or sheltered setting. They are both commonly considered to be agricultural pests and would likely represent opportunistic prey for the resident farmers. It is also possible that people at the site deliberately or inadvertently created prime habitat for pocket gopher and cottontail, disturbing the surrounding ground

Table 7. Plant and animal taxa identified at Kite Pithouse Village

Fauna (animal remains)	*Common Name*
Antilocapra americana	Pronghorn (American antelope)
Odocoileus spp.	Deer, probably mule deer
Canis cf. *Familiaris*	Domesticated dog
Sylvilagus audobonii	Desert cottontail
Lepus californicus	Black-tailed jack rabbit
Neotoma albigula	White-throated wood rat
Neotoma micropus	Southern Plains wood rat
Neotoma mexicana	Mexican wood rat
Cynomys ludovicianus	Black-tailed prairie dog
Spermophilus tridecemlineatus	Thirteen-lined ground squirrel
Spermophilus variagatus	Rock squirrel
Geomys bursarius	Plains pocket gopher
Peromyscus spp.	Deer mouse
Dipodomys spectabilis	Banner-tailed kangaroo rat
Dipodomys ordii	Ord's kangaroo rat
Flora (pollen remains)	*Common name*
Abies spp.	Fir
Juniperus spp.	Juniper
Picea spp.	Spruce
Pinus spp.	Pine
Quercus spp.	Oak
Salix spp.	Willow
Chenopodium spp. *Amaranth* spp.	Goosefoot, pigweed
Sarcobatus	Greasewood
Tidestomia	Tidestromia
Cleome	Bee weed
Compositae	Sunflower family
Artemesia	Sagebrush family
Cruciferae	Mustard family
Ephedra	Mormon tea
Eriogonum	Wild buckwheat
Eurporbia	Spurge
Geranium	Geranium
Gramineae	Grass family
Labiatae	Mint family
Leguminosae	Legume or pea family

Table 7. (continued)

Flora (pollen remains)	*Common Name*
Liliaceae	Lily family
Opuntia	Prickly pear cactus
Cylindropunti	Cholla cactus
Rhamnaceae	Buckthorn family
Rhus trilobata	Squawberry
Rosaceae	Rose family
Solanaceae	Potato/tomato family
Spaeralceae	Globe mallow
Zea spp.	Maize

Plant remains and pollen identified by Cummings (1988); fauna identified by Laura Elsinger and Holmes Semken (Elsinger and Semken1990); data from Rautman (1990).

with agricultural fields and storing vegetable foodstuffs. In either case, the smaller animal bone recovered at the site probably represents a fair sample of the animals that were locally available in the immediate site area.

The lithic remains from Kite Pithouse Village present a similar picture of reliance on locally available resources (Rautman 1990). The vast majority (nearly 85 percent) of the stone, including limestone, fine-grained chert and chalcedony of various colors, and also a handsome purple quartzite, could be obtained from the site area itself, or from within a few kilometers of the site. Areas of Quaternary pediment gravels mapped nearby record where coarser stone was left by flowing water during the Ice Ages (Bates et al. 1947; Brubacher 1982). These small relict gravel bars are insignificant for commercial gravel mining, but would have provided more than enough raw material for stone tool use. I visited one of these mapped pediment gravel areas and determined that the different cherts and quartzite recovered from the site could easily have been obtained from these local sources.

Among the varieties of chert used for stone tool manufacture, the single most abundant kind was an opaque white stone called Mountainair chert. Modern sources are visible today in roadcuts about midway between the Kite Pithouse Village and the town of Mountainair.

Use of more distant lithic sources was represented by a very small amount of sandstone, siltstone, lava, and obsidian. None of these rock types is consistent with local geology, and the presence of these materials represents procurement from some source an unknown but farther distance away. The closest lava sources are about eighty to ninety kilometers distant, near Los Lunas to the northwest or Carrizozo to the south (Chapin and Seager 1975). However, all of these non-local materials together made up only 3.6 percent of the lithic assemblage.

The local abundance of suitable raw materials was also manifest in an expedient strategy of stone tool production and use, but some formal tools were manufactured of local stone as well (Rautman 1990). In general, however, there was little apparent discrimination among the fine-grained materials at Kite Pithouse Village, suggesting that differences in the availability of raw material were not particularly important in structuring the assemblage.

The midden deposits also yielded abundant evidence of ceramic use and discard. Most of the pottery sherds came from brown ware types such as Jornada Brown ware and Corona Corrugated, and the vast majority of the decorated ware was Chupadero Black-on-White. This pottery type is also very common at early pueblo sites. Most Chupadero pottery occurs as large globular jars with a small cylindrical neck, with handles placed at the neck. Hemispherical Chupadero bowls of modest size are also a common vessel form. These are relatively small, the size of a modern cereal bowl, capable of holding less than one liter of liquid. Archaeologists generally assume that these small bowls represent use by individuals, as opposed to functioning as serving bowls for group events.

The variety of non-local pottery types present, however, indicated that the people of Kite Pithouse Village maintained social contacts and exchange relationships with many different groups across a large area. Regional comparisons of ceramic assemblages from other areas (Rautman 1993b) suggested that the inhabitants of Kite Pithouse Village likely maintained these spatially extensive social ties as one means of coping with fairly regular but unpredictable variation in local environmental conditions during this period of early farming. The ceramic information indicates that the site's inhabitants maintained some contacts with other groups by the Rio Grande, but that the majority of their social ties were with groups to the southeast, in the upland areas closer to the Sierra Blanca Mountains (Rautman 1993b, 1996; Rocek and Rautman 2007, 2012).

Evidence of Maize Dependence

The major contribution of Kite Pithouse Village to our understanding of early village society, however, stems from the fortuitous discovery of a small sample of seven human burials. These burials were not in the midden, but were found in two different pit structures. Three burials were found within Structure 2, a small, relatively shallow pit structure. Four additional burials were found within the fill of a larger, deeper pit structure (Structure 6) that was stratigraphically below an adobe surface structure. This skeletal sample was particularly important in providing evidence for significant maize consumption during the Pithouse period (Katzenberg and Kelley 1991). Comparison with a skeletal sample of comparable size from the later site of Pueblo de la Mesa (discussed in chapter 5) showed that in fact there was very little change in diet over the time period of early villages that is considered here (Colyer 1996).

Aspects of Village Organization in Salinas Pithouse Sites

Obviously, our knowledge of regional settlement patterns during the Pithouse period is next to non-existent, and our understanding of site organization, spatial layout, the temporal overlap (if any) in occupation between the different structures within a site, and other aspects of the built environment is also woefully lacking. We have only a general understanding of change over time within the Pithouse period, some limited excavation evidence that tells us about how people used local and distant resources, the kinds of structures and features that they constructed, and some very limited spatial information about different activity areas within one site. Given this information, however, what can we tell about the social dynamics of the groups that occupied these archaeological sites during the Pithouse period?

Evidence of Residence Group Size and Composition

The single most obvious evidence of group organization that we have for this time period comes from the evidence of structure size. We know that during the early period of pithouse use, structures were roughly 12 to 14 square meters in size, about the size of a small modern bedroom. In contrast, later pit structures are significantly smaller, about 3 to 4 square meters. (In comparison, the great kivas in the northern Southwest measure from 100 to 800 square meters in area [Cordell and McBrinn 2012:178]; 100 square meters is roughly the size of a standard two-bedroom apartment in Albuquerque today.)

One practical approach used to interpret pithouse sites in the Southwest relies on estimates of floor area to infer social group size. To take one recent example, Altschul and Huber (2000) estimate dwellings of less than 32.5 square meters would most likely house a small group such as a nuclear family. Dwellings between 32.5 and 45.5 square meters would likely house a larger nuclear family or a small extended family. In contrast, houses larger than 45.5 square meters may represent an extended family household. Smaller structures are known ethnographically, but are usually occupied by individuals.

These ethnographic and archaeological comparisons suggest that the pit structures observed at all of the Salinas pithouse villages were likely occupied by small groups, around the size of what we might think of as a nuclear family—although the observed floor area of five to fourteen square meters is small according to this scale. Early-phase Salinas pit structures at the Pipeline site (and Structure 1 at Kite Pithouse Village) are roughly three to four meters in diameter, well within the reported range of Basketmaker III pithouses elsewhere in the Southwest (e.g., articles in Reed 2000).

Test excavations on the Gran Quivira property show that these later pit structures are about half that size (about 2.5 meters in diameter or about 5 square meters in area). This size is similar to that measured for Structure 2 at Kite Pithouse Village. It is possible that the occupation at Kite Pithouse Village spans both the early and the late portions of the Pithouse period, with the deeper, larger structures (Structures 1 and 6) representing the earlier occupation, and the shallower, smaller structure (Structure 2) representing the later occupation.

This interpretation, however, relies only on architectural size and form, and in fact the details of Structure 2's construction and original internal features may have been disturbed by its secondary use for burials. The chronometric dates, however, indicate that the midden at Kite Pithouse Village dates to the later Pithouse period; this assessment also agrees with the ceramic data available.

In this study, however, many of these ideas about group composition and social organization cannot be investigated with the information available from the very limited excavations at two sites. With these caveats, therefore, I rely primarily on evidence of structure size, while acknowledging that residential group size and composition are only loosely related concepts (e.g., Adams and Kasakoff 1976).

In sum, the size of pit structures across the entire Pithouse period is consistent with occupation by very small social units such as a nuclear family, or even a subsegment of a nuclear family. We do see a decrease in the size of residential structures over time, suggesting a change in the activities that took place within structures, or possibly in the timing or duration of structure occupation. At the same time, we also see a change in structure form (more square in shape) that hints at greater contact with Mogollon populations to the south. The implications of these observed changes for residential group size and composition, however, are not at all clear.

Evidence of Cooperative Economic Groups

If we assume that nuclear family groups represent the basic unit of household organization during this time period, can we tell anything more about the social relationships between these groups? In this section, I examine the evidence that is available from the spatial relationships between residential structures and storage features to assess what inferences we might be able to make from the small amount of information available.

We know that at the Pipeline site and also at Kite Pithouse Village, there is evidence for extramural as well as intramural storage pits. Flannery's (1972, 2002) studies of early village society in general argues that interior and exterior storage pits represent two different kinds of cooperative groups. His ethnographic data show that exterior storage pits are commonly associated with small, nuclear family households that cooperate with neighboring households to share labor and also foodstuffs.

The exterior position of the storage pits in this kind of settlement suggests that storage facilities were cooperatively owned, filled, and utilized (Kent 1999). Access to the contents of the pits would be visible to all, and likely a social event involving all concerned. Exterior storage pits are also commonly associated with groups that have a semisedentary settlement pattern in which people expect to vacate the site for extended periods of time, but then return (e.g., Kent 1984, 1991, 1992; Gilman 1987). At Kite Pithouse Village, the presence of one extramural storage pit exposed in the arroyo suggests a similar organization of at least seasonal sedentism at this site, with relatively long-term (at least seasonal) food storage.

The location of extramural storage pits in relation to residential structures is also an indicator of social relationships among the groups that would use these storage pits. Drawing from different ethnographic cases, there is an assumption that the occupants of any given residential structure would be able to monitor, and thus control, the extramural pits that were physically close (Kent 1984). Accordingly, some Southwestern archaeologists distinguish between those extramural pits that appear to be associated with a particular residential structure and those that are grouped together with other storage pits, away from the residential structures (e.g., see examples in Geib and Spurr 2000:185). These distinctions are usually interpreted as representing individual household or whole group control over stored items (e.g., Byrd 1994; Flannery 2002).

Appreciating that the Pipeline site sample comes from a single long test trench, there does seem to be some spatial clustering of the storage pit features within fifteen meters of each other on the eastern portion of the site. One of the excavated residential structures (Feature 5) was located some ninety-one meters farther west (Fenenga 1956). It is not clear where the other excavated residential structure (Green 1955) was located along this line. We also do not have any information about how this trench line relates spatially to the other slight depressions that Fenenga (1956) thought might represent other residences.

As a result, this small bit of information of site organization may not be particularly significant, but it does indicate that the storage pit features were not necessarily in close proximity to the residential structures, as we might expect if each nuclear family occupied one residential structure and controlled access to one or more associated features. Rather, this apparent clustering of storage pit features suggests that they were located in the open, and were thus possibly considered property of the larger group, rather than any one particular nuclear family.

At Kite Pithouse Village, there simply is not enough information to make any sort of inference about the spatial position of residential structures to storage pit features. The one large bell-shaped storage pit was located some distance away from the observed pit structure residences. We also know that Structure 1 had three interior storage pits (Beckett and Wiseman 1982). Structure 2, which we excavated in 1986, also showed at least one intramural pit, as well as three extramural pits that were immediately adjacent to the presumed residential structure (Rautman 1990:figure 1.7). However, the intramural pit (Feature 19) contained a burial, as did two of the three extramural pits (Features 9 and 14), so it is not at all clear whether these pits were dug for the purposes of the burials or were originally storage pits that were modified into burial pits. As a result of this potential remodeling for burials, the association between individual residential structures and extramural or even intramural storage pits is inconclusive at best.

Another line of evidence regarding the economic relationships between residential units comes from an examination of the size of the known storage features. Again, based on observations from ethnographically studied societies, archaeologists generally assume that the size of the storage feature has some consistent relationship with the size of the group that shares the rights and responsibilities involved in the use of its contents (e.g., Christakis 1999; Wesson 1999; Ames et al. 2008). Thus, small intramural pits are usually associated with a relatively small resident group, while very large communal storage rooms or buildings are associated with village-level cooperative groups.

In this case, the Pipeline site does provide some significant information regarding the size of the observed extramural bell-shaped pits and the range of variation present among them. These pits are very large indeed, roughly 1 to 1.5 meters in diameter and of similar height. Those at the Pipeline site show evidence of interior burning; the one exposed in the arroyo at Kite Pithouse Village shows a thin clay lining. Both kinds of treatment suggest some sort of planned long-term storage of perishable items. Even without knowing the spatial relationships between different pithouses and these features, or the continuity of occupation at each site, we can infer that the sites' inhabitants stockpiled food or other perishable goods for use at some later time (DeBoer 1988).

Estimating the storage volume of the bell-shaped storage pits can provide at least some minimal idea of the size of the local group that contributed to this stockpile. Here, numerous other studies contribute to the different lines of evidence used in calculating group size from storage capacity. In the following discussion, I use the estimates presented by Altschul and Huber (2000:152), who also outline the logic involved in each step of the calculation.

Calculating storage volume of the bell-shaped pits as a simplified cylinder results in an estimated capacity of a maximum of 2.6 cubic meters in volume, but the actual storage volume of the bell-shaped pit would be somewhat less because of the constricted mouth. This storage volume is slightly less than that of the 3 cubic meters that Altschul and Huber (2000) report for exterior storage bins for Basketmaker III period sites in Arizona. Altschul and Huber (2000:152) use information from estimated crop yields and caloric requirements to argue that each of the 3-cubic-meter bins at their archaeological site could have supplied a family of five with enough shelled corn to provide calories for 1.8 years. They point out, however, that this estimate assumes that people were eating only maize. A more realistic dietary reconstruction assumes that maize consumption represented about 60 percent of the diet. In this case, they argue, each bin (or pit) would provide storage capacity to supply one family for about three years. Based on historically documented crop yields from Navajo dry farming in Lukachukai, Arizona, they further estimate that same storage capacity would suffice to hold the shelled corn from nine acres of cornfields (Altschul and Huber 2000:153). As a comparative measure, in 1934–1935, Navajo farmers planted about six acres of corn per nuclear family (Hill 1938, cited in Altschul and Huber 2000:152).

Using these estimates to evaluate population sizes during the Pithouse period in Salinas is difficult, however, because we still lack basic information regarding the number of residential pit structures or storage features at any of the Pithouse period sites. We also do not really know what was stored in these bell-shaped pits. However, we do know that maize was abundant at the pithouse sites, and from the human bone chemistry, we know that maize was a significant part of the inhabitants' diet. Thus, despite the problems noted above, we can still make some fairly reasonable generalizations.

First, based on observations of the labor demands for dry farming maize among the Navajo, it seems reasonable to say that one large storage pit, such as those observed on the Pipeline site, could have held the planting and harvesting efforts of at least one nuclear family. Second, the scale of production postulated here would not necessarily involve economic cooperation between nuclear families in any formal or sustained arrangement, as we might see, for example, in societies with community granaries (e.g., Byrd 1994).

Third, each of these observed extramural pits were roughly similar in size. There is no evidence that any one storage pit–using group was particularly larger or more industrious than any other group. Together, these lines of evidence suggest a social group of multiple nuclear families that may have cooperated with one another on a fairly informal or as-needed basis, but without any evidence of formalized decision-making structure or internal social differentiation.

Evidence of Site Planning

The spatial organization of structures and features on each site is also evidence of occupation continuity and duration. In general, archaeologists seem to agree that the larger the resident group, and the more continuity and length of occupation, the more potential there is for spatial patterning of activities within the site (Kent 1984). This observation doesn't imply that early villagers were necessarily consciously concerned with site planning or the spatial separation of different activities. It does mean, however, that to the extent that residents shared assumptions about the appropriate context and location for different activities, we can expect to see the cumulative effect of their repeated actions as a patterned use of space within an archaeological site (Scarborough 1989; Crown and Kohler 1994; Altschul and Huber 2000).

At Kite Pithouse Village, the observed accumulation of thick extramural midden deposits is one such line of evidence that indicates some degree of social agreement about the appropriate use of different areas within a site. Such spatial separation of waste disposal from living areas argues for a relatively formal spatial organization of activities within the site itself, and the thickness of the midden testifies to the continuity of this spatial segregation over time (Varien and Ortman 2005). This kind of long-term spatial patterning suggests the kind of behavior that we might expect to see in a residential site where people consistently and readily identify the designated activity areas, even across the passage of some years or generations. It seems likely, therefore, that even if people used the site only seasonally, they not only expected to return but also did so regularly for some time.

This kind of information about site spatial organization simply does not exist for the Pipeline site. We do know, however, that at least two residential structures were reused for midden disposal by later inhabitants of the area, suggesting that the site was occupied for more than the expected use-life of any one structure.

Evidence of Social Commitment to Place

This observation contributes also to our ability to infer something about the relationship between the local group and the village site itself. The various structures and features themselves may represent fairly short-term occupations. From experimental archaeology and also from accumulations research in the northern Southwest, we know that the estimated use-life of a residential pit structure is about ten to twenty years (Cameron 1990; Varien and Ortman 2005; Crow Canyon Archaeological Center 2011).

However, at both Kite Pithouse Village and at the Pipeline site, there is evidence that site occupants were aware of, and participating in, maintaining the continuity of the site as a whole for a time frame longer than that represented by the use of an individual structure. For example, simply the fact that the residential pit structures themselves were filled with midden deposits shows continuity in the meaning that people attached to this locale as a place. Structures no longer used for habitation were not filled with windblown sand, but were repurposed as a locale for discard of household debris, resulting in a dense midden deposit with an abundance of ceramics, lithics, animal bone, and charcoal (Green 1955; Fenenga 1956; Rautman 1990). At Kite Pithouse Village, it is also significant that pithouses were also used for burials, presumably after they ceased to be used as habitation structures.

The existence of midden deposits within these structures supports the argument for relatively long-term occupation of the general site area, long enough for these given structures to be repurposed for midden disposal by later inhabitants. At Kite Pithouse Village, too, we see adobe surface structures placed directly atop an earlier pit structure, suggesting that the site occupation here likely spans many generations. These simple observations provide another line of support for the idea that Salinas Pithouse period groups showed considerable consistency in their return and reuse of selected site locales.

A related observation comes from the fill within the observed extramural storage pits. At the Pipeline site, Fenenga (1956) described four large storage pits that were exposed in the pipeline trench (Features 2, 3, 4, and 6). One of these storage pit features (Feature 4) represents two intersecting pits, which he interpreted as representing two different constructions, rather than a later enlargement of the original pit. He notes that all of the storage pit features were filled with "midden" or else "mixed midden soil" (Fenenga 1956:227, 229). Fenenga's choice of words apparently refers to the presence of charcoal flecks within the feature fill, because he notes artifacts separately. Only a very small number of sherds were found in three of the features.

This general lack of artifacts in the storage pit fill suggests that they were used until the end of the occupation of the site as a whole, and were left open to fill with windblown sand and debris after site abandonment.

Only one large extramural storage pit was identified in the arroyo at Kite Pithouse Village, and it was not excavated. Field observations showed, however, that this feature was filled with charcoal-stained sand, and no artifacts were found in the sediment or pollen samples taken from the fill. Although this observation concerns only one bell-shaped storage feature, it indicates a degree of continuity of occupation at the site locale, and also that some structures were abandoned before others. The people who continued to live in some portion of the site used these older, unoccupied residential structures for discard of daily refuse, including the debris of everyday life such as potsherds, lithic flakes, and animal bone, along with charcoal.

Pithouse Village Communities in Salinas

So does Kite Pithouse Village really represent a village? In fact, I think so, at least in some respects. That is, the Kite Pithouse Village and also the earlier Pipeline site both exhibit certain aspects of the built environment that we associate with life in an early village community; they display at least some of the characteristics of early village society.

Different aspects of the built environment described above include observations of structure size and shape, the sizes of storage features, the location of these structures and features within a given area, and other evidence of site spatial organization. These observations allow us to infer certain specific characteristics of the social groups inhabiting these archaeological site locales and the dynamics of their interactions. We have some idea that the minimal social group size was likely a nuclear family unit; we have some idea that different coresident families are engaged in farming and perhaps cooperating with other families for at least some tasks, and perhaps even pooling some of their resources. More cautiously, we might say that there is no evidence of tight residential group control over stored resources, as we might infer from a site exhibiting only intramural storage pits. We have a sense of these different families being relatively similar social units, with no evidence of social differentiation appearing in the archaeological record.

Just as a comparative reference, Wills and Windes (1989) estimate that Basketmaker III pithouse settlements in Chaco Canyon that date to about A.D. 500 to 700 would have been occupied by one to three nuclear families at any given time. I would hesitate to attempt a specific number on the expected momentary population sizes involved at the Pipeline site or at Kite Pithouse Village, but the thick and dense midden at Kite Pithouse Village seems to suggest a somewhat greater momentary population size, as well as considerable continuity of occupation over time.

We certainly get a sense of a certain longstanding commitment to a particular locale, or sense of place on the landscape: that is, a place where people either reside permanently or return to with some regularity (e.g., Schlanger 1992; Seymour 2009). We get the idea that this sense of place continues for a considerable period of time, perhaps generations. It also seems clear that the village site has a planned or at least a customary pattern to its spatial organization, with specific areas of the site reserved for midden disposal (at the Kite site, at least). There does not seem to be much emphasis placed on village social integrative structures, such as specialized communal buildings. On the other hand, the simple act of communal contribution to a shared midden space may also serve to convey some concept of social integration. All these observations do suggest that the people at Kite Pithouse Village did share a certain sense of a persistent group identity that was larger than the nuclear family unit and some shared assumptions about spatial organization of activities within a given locale, and also exhibit some degree of long-term connection with a meaningful place in the natural and cultural landscape of Salinas (see Varien 2002 for another example of this line of reasoning).

On the other hand, of course, there is much that we do not know about the internal dynamics of group life for the inhabitants of these pithouse sites. There is some indication of change over time in the size of the residential group, but we really do not yet understand what that means for the dynamics of social relationships between families. We also do not know whether our village sites represent a few families, or many, or whether use-rights to this area were broadly shared, or reserved only to group members.

In sum, there is a lot that we do not know about early pithouse village communities in the Salinas area. This review of a small amount of information from limited excavations at different sites stretches the available data quite far, perhaps too far. Southwestern archaeologists who are used to working with enormous databases derived from extensive excavations of dozens of structures from pithouse sites may find it hard to believe that we can neither identify the sequence of occupation of each pit structure on each site nor compare their relative sizes, shapes, intact floor assemblages, and spatial relationships across large areas of a site. Nevertheless, the limited data that we do have are important in establishing some of our baseline knowledge regarding the Pithouse period and in providing some comparative information for understanding later periods of occupation. And with this comparative framework in place, we can investigate what aspects of life changed in the succeeding Jacal period.

CHAPTER THREE

Jacal Village Sites and Communities

The early part of the Pithouse period in the Salinas Province coincides with late Basketmaker III occupations in other areas of the American Southwest, those dating after about A.D. 500 in the northern Southwest (see table 2). In many respects, events in the Salinas Province are considered to be "late" in comparison with cultural traditions in the northern Southwest, and one way that this time lag is manifest is in the architectural record of the earliest village sites. This situation can lead to some confusion if one expects the terminology of the Pecos chronological system to coincide with architectural changes in the Salinas area. As table 2 (pp. 32–35) shows, later pithouse sites such as Kite Pithouse Village are "Pueblo I" in date, and the earliest aboveground structures in Salinas span from the latter part of the Pueblo II period (A.D. 900 to 1150) into the Pueblo III period (A.D. 1150 to 1300). In this geographic area, therefore, archaeologists may use the Pecos terminology, but only in this comparative sense of situating local events into the larger Southwestern framework (e.g., the chapters within Spielmann 1998c). Here, I use primarily the local chronological system defined by Caperton (1981), which is based on observed changes in architecture from pithouse to jacal structures, and then to adobe and masonry pueblos.

The Jacal Period in Salinas

The earliest aboveground structures in the Salinas region are made of wattle and daub, a kind of construction that is locally called jacal. After the wood and adobe mud disintegrate and wash away, jacal structures appear as a line of partly buried upright tabular stone slabs, nearly touching each other, extending about ten to thirty centimeters above the ground surface.

Sites from this time period are known from some reconnaissance survey (Caperton 1981) and from Matthew Chamberlin's fieldwork that included archaeological survey in the area southwest of Gran Quivira (Chamberlin 2008). Table 8 lists the Jacal period sites known in Salinas, and their locations are shown in figure 11. No sites from this time period have excavation data. In this overview, I augment the available published survey information with other studies from adjacent areas such as the lower Rio Grande valley (Marshall and Walt 1984).

Identifying Jacal Sites in Salinas

Jacals are structures of one or more rooms that were constructed with single or double rows of upright stone slabs that project above the ground surface. At the time of use, however, the superstructure included upright wooden posts set into the ground adjacent to the slabs. In the case of the double rows of stone, the posts would fit between the rows. The whole wooden jacal superstructure formed the mesh for adobe plaster, which was either slathered directly over close-set posts or smeared over a basket-like framework of smaller interwoven sticks between the uprights. Although the vertical foundation stones are set into the ground, and presumably provided some stability for the wooden posts, they are not really foundation stones in the sense of protecting the wood from direct contact with the soil. Rather, archaeologists seem to assume that the slabs stabilized the bottoms of the posts and protected the adobe walls from surface runoff and erosion (Stuart 2000:53).

Around Gran Quivira, a different kind of jacal construction forms what Caperton (1981:6) called "transitional structures," which apparently date to about the last fifty years of the Jacal period, shortly before A.D. 1300 (Caperton 1981; Chamberlin 2008). These aboveground structures show more extensive use of larger masonry blocks for wall footings, with the higher portions of the walls consisting of jacal or even an adobe and stone mixture. One might think that upright slabs would not provide a very stable foundation for building adobe or stone walls above. Nevertheless, this practice was obviously of some importance for some reason—one can see these upright slabs used in later pueblo structures, and also lining the interior of several kivas at Gran Quivira.

Most of our information about jacal sites comes from the western portion of the Salinas area, in the uplands of Chupadera Mesa. As far as we

Table 8. Jacal Sites in the Salinas Province

Site Number	*Comments*
Northern area (Tompiro Division)	
LA 9004	Five room blocks
LA 9005	Two room blocks, 9 rooms and 2 rooms
LA 9010	Two room blocks
LA 9011	No data
Southern Area (Jumanos Division)	
LA 197j	Jacal site near Montezuma Pueblo (Chamberlin 2008)
LA 2004	Jacal site, possibly earliest in date (Chamberlin 2008:87)
LA 9013	Jacal site; south of Turkey Ridge (Caperton 1981)
LA 9014j	Jacal site near pueblo on Turkey Ridge (Chamberlin 2008)
LA 9015	Highly nucleated jacal site on Turkey Ridge; no associated pueblo (Chamberlin 2008)
LA 9016j	Jacal site near pueblo on Turkey Ridge (Chamberlin 2008)
LA 9017	Jacal site; Caperton (1981) reports possible pithouses
LA 9022	Jacal site; west of Turkey Ridge
LA 9023	Jacal site; west of Turkey Ridge
LA 9024	Jacal site; west of Turkey Ridge
LA 9025	Jacal site; east of Gran Quivira Pueblo
LA 9026j	Jacal site near pueblo (Chamberlin 2008)
LA 9030	Jacal site; north of Turkey Ridge
LA 9031	Jacal site; north of Turkey Ridge
LA 9032j	Jacal site near Frank's Pueblo (Chamberlin 2008)

Note: Data compiled from Caperton (1981), Stuart and Gauthier (1981); updated with information from Chamberlin (2008). Piro Division (Mera 1940) sites on Chupadera Mesa and its western slope are not shown here (see Caperton 1981). The "j" suffix denotes a jacal component near a pueblo site.

know, there are no jacal sites recorded around Jumanes Mesa and only one in the Medanos Plains. This apparent absence of jacal sites may indicate that jacal sites are in fact not there, or that no one has really looked for them, or that people who looked did not look in the right places or for the right surface evidence. Possibly the remains of these surface structures are buried under windblown sand.

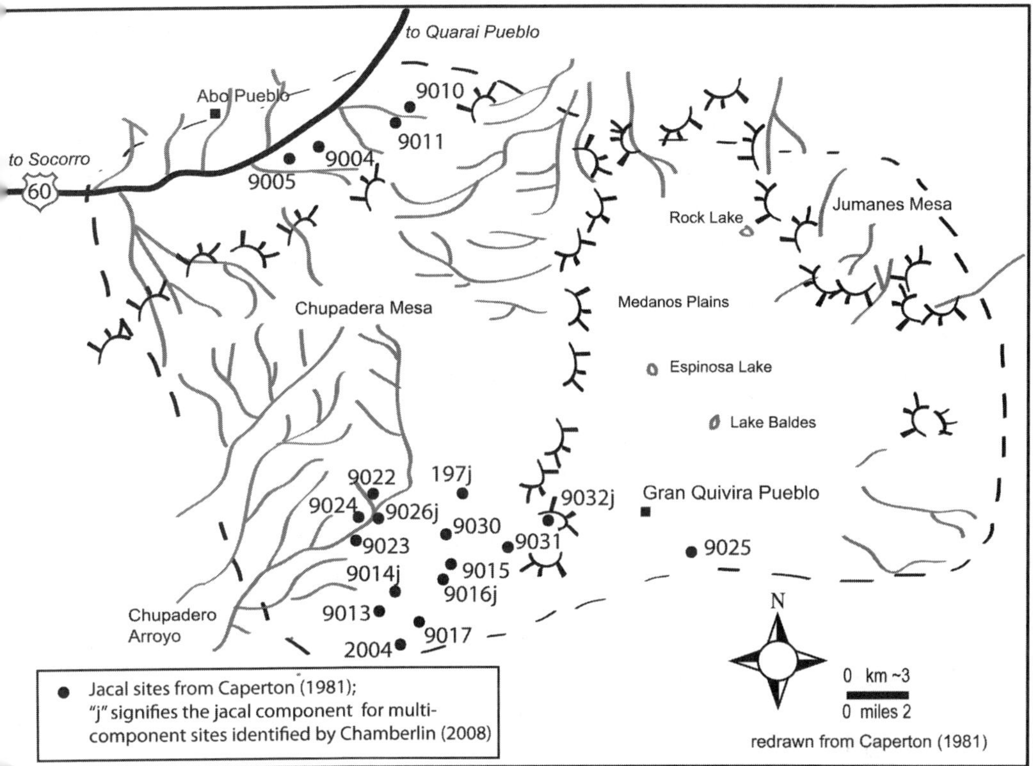

Figure 11. Location of jacal sites in the Salinas Province. Modified from the base map shown in Caperton 1981:xii.

Given the historic descriptions of soil erosion from agricultural fields near Gran Quivira, with sand dunes described as piling up against fences and buildings and blocking passage along the dirt roads (Link and Link 1999), such a scenario is quite reasonable. Alternatively, as Caperton (1981) reports, it is also widely assumed that jacal sites in the sandy lowlands were destroyed in the homesteading period as pinto bean farmers cleared rocks from their fields or scattered them by plowing. At this point, therefore, we really do not know how many Jacal period sites there might have been in the lowlands across the entire Salinas region.

The Known Jacal Sites in Salinas

Toulouse and Stephenson (1960) defined two different kinds of jacal sites, which they termed the Claunch Focus and the Arroyo Seco Focus, both spanning the time from about A.D. 1200 to 1300. These terms are not used today, but their scheme is helpful in pointing out some important similarities and differences among the early sites within the region. In their view, the Claunch Focus represents use of lowland areas. The sites include small jacal room blocks, each one to three rooms long. Typical ceramic assemblages on these sites include Jornada Brown ware, Los Lunas Smudged, San Francisco Red ware, Chupadero Black-on-White, and Corona Corrugated. In contrast, the Arroyo Seco Focus sites tend to be larger, plaza-oriented sites located in the uplands of Chupadera Mesa. Chupadero Black-on-White and Corona Corrugated are the dominant pottery types, while St. John's Polychrome is also present (Toulouse and Stephenson 1960:41).

Stuart and Gauthier (1981:322–323) disagree with this lowland/upland settlement dichotomy. They suggest that the larger Arroyo Seco Focus sites date to slightly later in time. The overall lack of information about Jacal period sites in lowland settings, however, makes it difficult to evaluate the significance of the observed variation in site location.

Caperton's reconnaissance survey (1981) and Beckett's (1981) survey within the National Park Service boundaries at Gran Quivira show that surface structures are sometimes found near pithouse architecture, and there is some evidence that the surface structures were in fact built later. At Kite Pithouse Village, for example, the excavated surface structure overlies a deep pit structure. Based on ceramic dates, however, Caperton concludes that there may be an overlap in the time of occupation of the surface structures with pit structures (Caperton 1981:5–6).

An additional complication is that some of the surface structures at Gran Quivira (Beckett 1981) and all of those noted at Kite Pithouse Village (Rautman 1990) are adobe-walled structures, without the upright slabs that are more typical of Jacal period sites. It is therefore difficult to know if these surface structures really should be considered Jacal period in date or if they may have been surface storage rooms associated with a late pithouse occupation. Caperton's (1981:6) assessment is that there is no Jacal period occupation in the lowlands around Gran Quivira, adding that nearby Kite Pueblo is the only known pueblo site that does not show evidence of earlier jacal structures.

Additional excavation data from the observed adobe surface structures is unlikely to resolve this issue, since surface structures such as those at Kite Pithouse Village were filled with aeolian sand and little else. At this point, we simply do not have enough information to establish their relationship to Jacal period occupations.

More typically, jacal structures occur in the same general area as later masonry pueblos. This spatial proximity again makes it difficult to distinguish artifact assemblages that are unique to each time period. As a result of this problem, Caperton's (1981) temporal estimates are based on the ceramic assemblages found at single component jacal sites. He found three single component jacal sites near Abó Canyon in the northern portion of the Salinas area, two sites near the (ghost) town of Chupadero on top of the mesa, and ten in the southern portion of Chupadera Mesa. For this reason, we know the most about Jacal period occupation in the southern portion of Salinas, near Gran Quivira. These single component jacal sites show that jacal construction coincides with other changes in material culture at about A.D. 1100, including the appearance of a new utility ware, Corona Corrugated, which replaces Jornada Brown ware, and also widespread use of Chupadero Black-on-White (Caperton 1981:6).

The second major source of information about the Jacal period occupation in Salinas comes from Chamberlin's archaeological survey of Turkey Ridge, a small upland landform southwest of Gran Quivira (Chamberlin 2001, 2006, 2008). He found that each of the known masonry pueblo sites included a dispersed collection of presumably earlier jacal room blocks scattered in the same general vicinity of the later pueblo. In addition, Spielmann (1989) also conducted a survey to the east of Chupadera Mesa, providing additional information about dispersed non-architectural sites in the Medanos Plains.

Jacal period sites on Chupadera Mesa have been dated primarily through evaluation of the ceramic assemblage (Chamberlin 2008). He distinguishes one jacal site (LA 2004) as the earliest in date, based on the predominance of Jornada Brown ware (67 percent of sampled ceramics). The ceramic assemblages at later jacal sites include lesser amounts of Jornada Brown ware (2 to 28 percent of the assemblage) and correspondingly greater amounts of Corona Corrugated. Jacal sites also show surface artifact scatters of Chupadero Black-on-White with minor amounts of White Mountain Red ware. In general, however, these ceramic assemblages date the identified jacal sites to the late twelfth or early thirteenth centuries. Aside from the one early site (LA 2004), the rest of the jacal sites seem to have been established at about the same time.

The lack of excavation data means that we have little information regarding the subsistence economy or other aspects of daily life during this time. Chamberlin (2008) does provide some information about ceramic production and exchange within the region, derived from neutron activation analysis of surface ceramics. Given the small amount of data available, this discussion of Jacal period community organization will include some information about Jacal period site organization from a survey that was done previously on the west side of Chupadera Mesa (Kyte 1988, 1989; Montgomery and Bowman 1989), and also a second survey of the lower reaches of the Rio Grande, the Rio Abajo (Marshall and Walt 1984).

The Rio Abajo is considered to represent Ancestral Piro occupation (Marshall and Walt 1984). We know that the Tompiro language, spoken within the historic pueblos of Abó and Gran Quivira, is related to Piro, so it might seem logical to assume a clear social relationship between people who lived in these two areas. In fact, however, the relationship between these groups along the Rio Grande and the early Salinas villagers in antiquity is not at all certain. Differences in Jacal period site organization between these two areas highlights some significant differences in community organization at this time of early villages and indicates significant local differentiation in community organization and village life.

Site Organization in Salinas Jacal Sites

Within the Salinas area, there is enough variation in jacal sites that Caperton (1981) distinguished two different overall patterns. In the northern portion of the Salinas region, near the pueblo and Spanish mission of Abó, the few known jacal villages are smaller, with only a couple small room blocks per village site. Most of the Jacal period occupation seems to be focused on the southern portions of the area, near Gran Quivira. Here there are more jacal sites, and the sites are larger. The individual rooms are of similar size, but the room blocks are generally longer, with more rooms per room block. There are also more room blocks per site. On sites where there is more than one room block, the different room blocks may be spatially dispersed across a larger site area; other sites show a more clustered or nucleated arrangement.

In the area around Gran Quivira, jacal sites generally are composed of one or more square rooms, each room about four square meters in size (two by two meters). Single rooms are rare; more commonly, rooms are organized in a room block of at least two rooms. In general, room blocks consist of a line of rooms, with a preferred orientation where the long axis is aligned approximately north-south. The length of the room block is variable, but typically ranges from two to five rooms. Longer room blocks are not uncommon, however, and some are up to a dozen or so rooms in length.

Some of the individual room blocks on Salinas jacal sites have linear extensions built out to the east. Depending on the number of these extensions and their location along the north-south spine, one can describe the whole room block as having the shape of capital letters L, and more rarely the letters I, F, E, C, and T (Caperton 1981). More rarely, a room block may be two rooms wide and of variable length. The average number of rooms per room block ranges from four to eight, encompassing an average roofed area of five to ten square meters (Chamberlin 2008:figure 4.10).

In some sites there are one or two such room blocks; in other sites there might be several room blocks loosely clustered within what people usually call a "village" or sometimes a "cluster," that is, a spatial grouping. The smaller jacal villages (clusters of room blocks in a given locality) may include about a dozen different contiguous room block structures; the largest one (LA 197j) showed 51 multiroom structures, totaling some 276 individual rooms (Chamberlin 2008:table 4.4).

The jacal sites also exhibit variation in the degree of nucleation or spatial clustering of the room blocks vis-à-vis one another. Chamberlin (2008) found that the earliest identified jacal site south of Turkey Ridge also exhibits the most dispersed site plan. He noted an overall trend toward greater nucleation over time, with the sites on Turkey Ridge being the most highly nucleated. This observed degree of nucleation is not, as one might think, dictated by the availability of open space on the ridge top, since there appears to be plenty of land available for a more sprawling village layout, even on the ridge top. Instead, he attributed this settlement trend to social or cultural factors (Chamberlin 2008).

Caperton (1981) reports that there is no apparent overall village plan or spatial organization of room blocks within a site. However, at a minimum, it seems clear that the long axes of Salinas room blocks do tend to be oriented north-south. In addition, Chamberlin (2008) also identified some degree of spatial organization at three different sites. One site, LA 9015, is highly nucleated, with several room blocks organized in a linear site plan. This arrangement left a cleared space to the side that could have accommodated larger group gatherings (figure 12). Chamberlin also identified two other sites with a cleared central space that could have been used as a plaza. This open plaza-like space was apparent at LA 9026j, and at the jacal component of a masonry pueblo site on Turkey Ridge, LA 9014j (Chamberlin 2008, 2012).

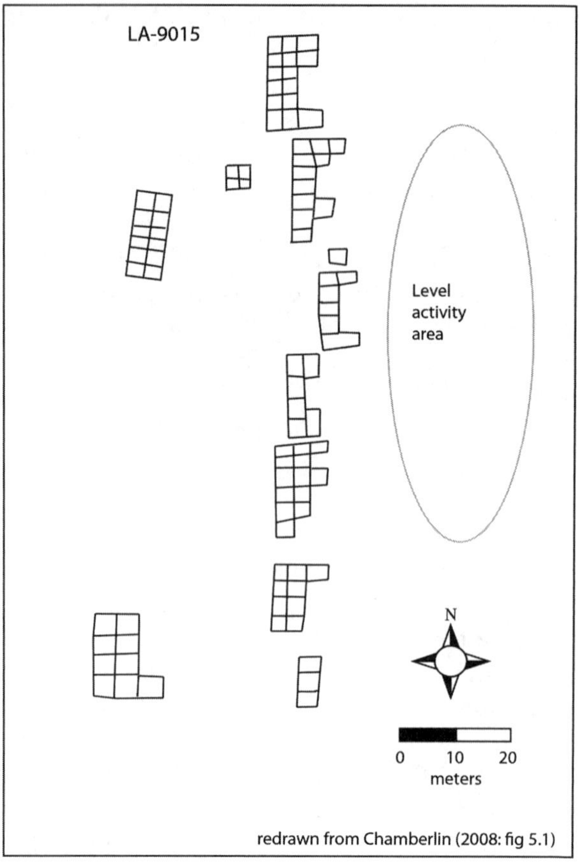

Figure 12. Jacal site with linear organization. Redrawn from Chamberlin 2008:figure 5.1.

Community Organization in Salinas Jacal Sites

Given this limited information about the location and layout of these jacal village sites, what can we say about the organization of these potential communities? As I mentioned for the Pithouse period, these inferences may not be as definitive or comprehensive as one would like, but they do provide some baseline information for comparison with other regions and with subsequent time periods in Salinas.

Evidence of Group Size and Composition

Residential room size itself is usually related to the size of the domestic group (see Wilk 1984; Stone 2000); cross-cultural studies suggest about ten square meters per person is common (Naroll 1962). Smaller rooms, such as the four-square-meter rooms common for jacal sites and in the later Salinas pueblos, are usually interpreted as non-residential storage rooms or as representing a portion of a nuclear family dwelling. While we lack information about the interconnections between rooms on jacal sites, it seems likely that individual households would have occupied more than one room. In contrast, large rooms such as those found in the Rio Abajo (see below) that are closer to twenty-five square meter rooms are usually interpreted as housing more people, such as a family grouping (Cameron 1999a).

In the Pueblo Indian Southwest, the number of rooms per structure (room block) is also indicative of the size of the residential group. For example, in the northern Southwest, early sites showed use of pit structures backed by a line or arc of contiguous single-story surface structures. In some of these cases, the pit structures were used as residences, and the surface structures were apparently used for storage (Cordell and McBrinn 2012). It is possible that the adobe surface structures that I mapped at Kite Pithouse Village (see chapter 2) represent this type of surface-room storage feature.

It is interesting that the typical room size observed for Jacal period sites (four square meters) is relatively consistent with the size of the small, late pit structures such as those mapped at Gran Quivira and at Kite Pithouse Village. As I mentioned in chapter 2, similar small residential pithouse structures are also known from nearby regions, for example, in sites near Belen closer to the Rio Grande (Ferdon and Reed 1950). The significance of this temporal continuity in room size between the later Pithouse period and the Jacal period suggests a similar continuity in basic social organizational structure (family grouping) across time.

Of course, one of the major differences in architecture between pithouses and pueblos is the contiguous arrangement of rooms within a surface pueblo. In the next section, I consider the implications of the basic concept of the contiguous room block for Salinas farmers of the Jacal period.

Evidence of Cooperative Groups

Whatever the relative advantages or disadvantages of pit structures compared to surface structures (e.g., Gilman 1987), in the Pueblo Southwest, the room block structure of contiguous surface rooms represents the most commonly used residential structure of later time periods. When considering room block architecture, however, several dimensions of variation are significant. First, there is the number of rooms included in each room block, as well as their spatial relationship to one another. Second, archaeologists also consider the number and arrangement of interconnected rooms, or room suites that may be apparent within a single room block. Third, there is also the issue of the square footage involved in individual rooms, in the room suites, and also in the roofed area of the structure as a whole.

In chapter 1, I referred to the Duckfoot site near Mesa Verde as an example of one of the larger early pueblo sites that date to about A.D. 850 (Lightfoot and Etzkorn 1993). This site includes several such multiroom groupings (suites) encompassed within one contiguous room block. At the Duckfoot site, each room suite within a single room block fronts on a single pit structure. In this particular case, the idea is that each room suite would have been occupied by a group of people, and the apparent variation in the number of rooms per suite (five to eight) indicates some variation in the number of individuals in these groupings.

The social interpretation of this architectural arrangement is that these larger groups may have cooperated as an economic unit, a household, in which members shared living space and interior storage facilities. Each extended family household would also presumably utilize or possibly control access to the "courtyard" area in front of the room block as well as the pit structure immediately adjacent (Varien and Lightfoot 1989; Lightfoot and Etzkorn 1993).

In the Salinas region, evidence from surveys indicates that single detached rooms were uncommon on Jacal period sites. Rather, jacal structures were more typically organized in groups of two or more rooms. Different room blocks vary in size, of course, but usually include less than a half-dozen rooms, arranged in a single line. At the jacal village site associated with LA 197 (Montezuma's Pueblo), for example, 59 percent of the room blocks had fewer than six rooms (Chamberlin 2008). Chamberlin's survey recorded only one site (LA 9026j) with significantly larger room block structures. At this site, one room block grouping had fifteen rooms and one had twenty rooms, representing the largest number of rooms observed within a single contiguous structure (Chamberlin 2008).

In contrast to the apparent internal organization of rooms within room blocks at the Duckfoot site, however, we have no evidence regarding doorways, access routes, or other forms of possible internal differentiation.

Without making any assumptions regarding the use of these rooms for sleeping or for storage, we can observe that the room itself (about four square meters) is the basic unit of the room block, and that room blocks consist of a simple iteration of this unit in a fairly straightforward manner.

Further, we know that most of the room block structures themselves form straight lines of contiguous structures; only the two largest room blocks at LA 9026j are two rooms across. No larger massing of rooms is apparent on the observed jacal sites. That is, all rooms would have had equal access to exterior space; there are no completely "internal" rooms that could be accessed only through other rooms.

Thus, while there is considerable standardization in room sizes, the observed variation in the number of rooms per room block does indicate considerable variation in the size of whatever social unit might occupy such a grouping. Chamberlin (2008:95) notes some specific dimensions of this variation. For example, the site described above, LA 9026j, exhibited the two largest room block structures in the study area, each two rooms "across." However, LA 9015, on Turkey Ridge, exhibits the longest set of linear contiguous rooms. These two types of room blocks presumably represent some sort of different organizational principles, but without knowledge of possible room interconnections, or intraroom activities, it is difficult to infer what those principles might be. The fact remains, however, that we do see some variation or experimentation in people's basic strategy of room block formation

To the extent that we can infer social units from the size of these contiguous room blocks, it is clear that Salinas jacal villagers were organized in relatively small groups. It is interesting to note that the typical or commonly observed room block, averaging six rooms or fewer, is not that far removed from the five to eight rooms included within a single suite of rooms at the Duckfoot site in Colorado.

We do not know if this hypothetical group was strictly a residential arrangement, or whether these groups might have represented an economically integrated household, or shared responsibility for household decision-making. Unlike at the Duckfoot site, there is no obvious shared workspace (courtyard) or pithouse/kiva structure spatially associated with this possible social unit. And the Salinas sites do not have excavation data. At this point, therefore, we will have to just note the observed similarity of commonly observed group size without making any necessary inferences about their internal organization or defined activities.

The range of variation observed in the number of rooms per room block, however, indicates a relatively high degree of variation in the proposed size of the residential group. This variation is seen within each village site (archaeological site), although of course the sites discussed above represent the largest range of variation among the observed sites. A certain amount of variation in the number and size of these proposed social groupings, therefore, represents a widespread regional pattern, with two jacal sites exhibiting a larger range of variation compared to other similar sites in the area.

There is also some indication of a change over time in the size of the largest social units, perhaps signifying increased intravillage variation over time. The only early Jacal period site that we know about is LA 2004. This site has an average of 1.4 rooms per room block. The most common arrangement, however, is a two-room structure, and no room block contained over six rooms (Chamberlin 2008). These data suggest that the most important social grouping at the site was very small indeed, along the lines of an individual family household. If this generalization is true, the later sites show some continuity of this basic organizational structure for many groups, but also a trend toward collaboration of these relatively small groups and formation of larger ones.

In the northern Southwest, evidence of social groupings larger than we might anticipate for a nuclear family is often interpreted as the development of kin-based lineages and clans. This interpretation, of course, assumes some degree of temporal continuity for the social organizational principles and groupings that are known from the ethnohistoric and modern Pueblo Indians (e.g., Carlyle et al. 2000; Peregrine 2001).

In this discussion of Salinas village communities, however, investigating the relationship between the observed residential room blocks and kinship organization is unlikely to be a productive line of research until we have considerably more information and excavation data from the Jacal period occupation. The important observation for this study is that during the Jacal period, these residential units, whatever their internal composition, are variable in size, but varying within certain given parameters that are consistent across the whole region. The differences that we do see between larger and smaller social groups within a village community may or may not have affected other aspects of life, such as economic well-being or social status, or access to valued goods and ideas.

Evidence of Village Integration and Cooperation

In contrast to the situation during the Pithouse period, where there is really no evidence whatsoever about village site spatial organization,

Chamberlin's survey provides exactly the kind of spatial data that contributes to an evaluation of the social processes that are important for creating a social community and for identifying the presence of these kinds of social interactions in the archaeological record. As noted in the introduction, some of the attributes of a functioning community involve economic activities, including the group's ability to cooperate and coordinate subsistence activities and make decisions regarding shared resources (Kolb and Snead 1997).

Communities also involve enough people and interaction and continuity of interaction among them to ensure not just biological reproduction but also social and cultural reproduction. This kind of organization typically involves the social integration of groups larger than the household (Johnson 1982). A village community also involves some degree of self-awareness of the group as a whole, with a sense of group membership, participation, history, and temporal continuity. In spatial terms, this self-awareness is often expressed in, and created by, structures of the built environment that contribute to whole-group activities. The evidence that we have for the Jacal period in Salinas provides an opportunity to assess these aspects of community organization for jacal sites.

Much of the evidence for whole-group identity and decision-making comes from evidence that people within each village site were aware of, and accommodating the presence of, other people, other structures, and the surrounding cultural landscape. As noted for the Pithouse period, the architecture itself cannot tell us whether the site represents a series of occupations by a very small social unit, or an occupation by a larger corporate group. However, some aspects of community organization are expressed in the built environment, including (1) development of relatively nucleated settlements within a given range of variation and (2) internal site organization that included shared space such as a courtyard or plaza.

In fact, all the southern jacal sites are fairly dispersed in their site organization, especially from the perspective of later pueblo occupations. While none of the jacal sites show the well-defined plazas that characterize later pueblos in this region, three sites show evidence of some sort of communal space that was apparently deliberately maintained near the residential structures. Two of the jacal sites, at LA 9014 and LA 9015, showed evidence of internal organization of buildings around a cleared central space. Another jacal settlement, LA 9015, includes a linear plaza area off to the side (Chamberlin 2008).

This concern for visible space, large enough to hold significant numbers of people, is seen within the most nucleated jacal site (LA 9015) as well as in one of the more spatially dispersed settlements (LA 9026). Thus, this concept of communal space represents a strategy of social interaction that is independent of an observed trend toward greater settlement nucleation over time.

Another dimension of variation within settlements is that observed between the sites on Chupadera Mesa proper and those on the small isolated knoll of Turkey Ridge. In general, the jacal sites on the mesa top (LA 197j, 9026j, and 9032j) showed more dispersed settlement organization, with numerous smaller social units. On Turkey Ridge, in contrast, settlement nucleation included larger room blocks with larger rooms and also tighter clustering of all these structures (Chamberlin 2008). These village sites thus likely included larger social units and a greater emphasis on integration among those units. The high degree of nucleation, in the apparent absence of any pressure for living space on the ridge, seems to indicate a strategy of integration that included special integrative structures (plazas) as well as spatial proximity in everyday activities. These Turkey Ridge jacal sites (LA 9014j, 9015, and 9016j) also exhibited more investment in construction, that is, more use of masonry for more substantial structures (Chamberlin 2008).

In this case, Chamberlin's (2008) study established that Jacal period villages in Salinas did indeed show a high degree of cooperation, communication, and organization within each village community, at a time well before the construction of the masonry pueblos. He concluded that the jacal sites in Salinas represent a range of variation in certain generally accepted strategies by which village communities organized themselves spatially within a locality and within the built environment of the village.

Evidence of Social Commitment to Place

One of the most significant findings of recent research in the Salinas area has been the excavation and survey data showing long-term investment in people's sense of place. Caperton (1981) suggested such long-term occupation of Salinas as a region, and this idea has been reaffirmed with each archaeological project. There is evidence of fairly continuous occupation in favored locales from the Pithouse period onward. The pueblos that are associated with jacal occupation were occupied well into the 1300s, testifying to the stability of the local cultural landscape and the persistence of place across centuries of occupation. In later time periods, many of the small pueblos in these long-settled areas are in fact abandoned in terms of daily use and residence; at the same time the few remaining pueblos such as Abó, Quaraí, and Gran Quivira grew dramatically larger in size. It appears, however, that this aggregation represents a reassortment of an existing population, rather than any large-scale immigration from outside the region. These later population movements will be discussed later, in chapter 7.

Evidence of Regional Integration

The similarities in site organization, the variables associated with the built environment, and the long-term investment in specific places on the landscape suggest that early farmers in the Salinas region did share quite a number of basic cultural assumptions. In addition to shared subsistence strategies and use of similar kinds of material items such as ceramics, the Salinas people also shared many of the same ideas about what constituted a village community and how people within them should interact. We see these shared assumptions through the evidence of the built environment (see Morgan 1965; Rapoport 1982, 1988). The variation seen in the jacal sites involves both the number of rooms per contiguous room block, the number of room blocks per sites, and the spatial distance between room blocks.

These kinds of shared assumptions between the different jacal village communities within the Salinas region also can be seen in the evidence of ceramic production and exchange. Chamberlin's (2008) study of regional exchange patterns for decorated pottery types show that people in the different jacal village sites were broadly similar in their access to, and use of, decorated pottery from different sources. There are some differences noted: for example, jacal sites on the east side of Chupadera Mesa seem to obtain more of their Chupadero Black-on-White vessels from manufacturing areas near Gran Quivira, while jacal sites on the western side of Chupadera Mesa show closer ties to clay source areas in Chupadera Arroyo (Chamberlin 2008).

These exchange relationships, however, seem to coincide with simple distance factors and transport costs. There is no indication of any particular trend or patterning in the different jacal villages' access to locally made pottery or clay sources on Chupadera Mesa itself. In addition, the different jacal villages seem to have participated more or less equally in ceramic trade with more distant sources, such as those groups that produced White Mountain Red ware. The architectural evidence and also the evidence of local and also non-local ceramic exchange points to a relatively high degree of uniformity among the several Salinas-area jacal sites in overall layout, organization, and social interactions, both within and also between the different village sites.

To appreciate the extent to which Salinas-area jacal villagers shared these ideas of community organization, we need some comparative information about neighboring areas. In the rest of this chapter, I use survey data from comparable time periods along the lower reaches of the Rio Grande: the Rio Abajo survey (Marshall and Walt 1984). Another regional survey of the area immediately west of Salinas, the Chupadero Arroyo survey (Kyte 1988, 1989; Montgomery and Bowman 1989) focused on mapping some of pueblo sites that were already known and therefore contributes only a small amount of information for regional comparisons within the Jacal period. There is no comparable ceramic sourcing information from either of these surveys, so the following discussion considers only the architectural evidence.

Regional Comparisons: The Rio Abajo Area

Marshall and Walt (1984) use the term Rio Abajo to apply to an area along the Rio Grande from the confluence of the Rio Puerco to the north, south to the historic Piro pueblo of Senecu (near the modern town of Socorro; see figure 1). In this area, jacal structures overlap in time with pithouses, and also with cobble and masonry surface structures. The time periods in this area are called the Tajo Phase (roughly comparable to the later Salinas Pithouse period) and the Early and Late Elmendorf Phases (roughly the same time as the Jacal and Transitional periods in Salinas). Table 9 compares the time periods in these two survey regions.

Tajo Phase Sites

Tajo Phase sites date to about A.D. 800 to 1000, which corresponds to the Pueblo I period in the northern Southwest (Marshall and Walt 1984:47). These sites are small, with one to ten surface rooms; nearly one-quarter of the sites have pithouses as well. These pithouses are quite large, 3.5 to 5 meters in diameter, a size similar to the large early pithouses in Salinas. Eighteen Tajo Phase sites with surface structures were recorded, but these structures did not have upright stone slabs. Instead, the builders sensibly used readily available rounded river cobbles from the Rio Grande or its tributaries. Some jacals were single rooms, but others were linear unit houses (room blocks) of contiguous rooms. The typical unit house

Table 9. Comparison of time periods in Salinas and the Rio Abajo

Approximate Dates	*Rio Abajo Region*	*Salinas Region*	*Pecos System*
20,000 to 5000 B.C.	Paleoindian	Paleoindian	Paleoindian
5000 to 200 B.C.	Archaic	Archaic	Archaic
200 B.C. to A.D. 400	San Marcial phase	Archaic	Basketmaker II
A.D. 400 to 800	San Marcial phase	Early Pithouse period	Basketmaker III
A.D. 800 to 950/1000	Tajo phase	Late Pithouse period	Pueblo I
A.D. 950/1000 to 1100	Early Elmendorf phase	Late Pithouse and Jacal period	Pueblo II
A.D. 1100 to 1300	Late Elmendorf phase	Jacal period and Early Pueblo period	Pueblo III

The data used in the first two columns of this table are taken from Marshall and Walt (1984:table 1.5). Marshall and Walt (1984) also correlate the Rio Abajo phases explicitly with Pecos system terminology.

includes two rooms, and the mean number of rooms per site is five rooms (Marshall and Walt 1984). In this respect, then, the Tajo Phase sites seem quite similar to the early jacal sites in Salinas.

There is a significant difference, however, in the sizes of these rooms. In contrast to the Salinas area where room size is relatively standardized within a site, the Tajo Phase room sizes are highly variable—not only between sites but also within a single site or even within a single room block. Room sizes in the Rio Abajo sites are typically 4 to 13 square meters, but can be as small as 3 square meters and as large as 25 square meters. The largest rooms represent 5 x 5 meter rooms: a sizable open space (Marshall and Walt 1984).

Tajo Phase room block structures, like their Salinas counterparts, are generally linear in form. It is clear, however, that the Tajo Phase surface room blocks were commonly built by accretion—that is, with one room added against its neighbor (Marshall and Walt 1984). It is not clear whether the Salinas jacal sites show a similar accretional building strategy, but the overall greater uniformity of room size and construction methods that people used within a given site seems to indicate a higher degree of conformity and planning within each village community in the Salinas region.

The riverside Tajo Phase sites cluster spatially in two different areas, along the Rio Grande near the confluence with two different tributary rivers. There are some upland sites also, but they are smaller and more dispersed across the landscape. In this case, the Tajo Phase apparently represents a shift in settlement pattern from the uplands to riverside locations. Marshall and Walt (1984) report that the sites within each cluster are spatially fairly dispersed. For the observed groups of sites, they use the term "open community clusters," describing them as scattered hamlets of small room-block groups, each grouping about one hundred meters apart, but usually within sight of each other.

Early Elmendorf Phase Sites

Early Elmendorf Phase sites date to Pueblo II, about A.D. 950 to 1100 (Marshall and Walt 1984:75). In the Salinas area, this time frame spans the time of occupation of Kite Pithouse Village and also the Jacal period. Marshall and Walt (1984) report that this period records population movement from north to south within the Rio Abajo region, and the earlier Tajo Phase sites are mostly abandoned. Instead, an estimated 80 to 90 percent of the regional population aggregated into five newly established "village areas."

These Early Elmendorf settlements clearly sort into large and small sites. The small sites have one to nine rooms with a mean of four and a half rooms per site. Larger sites are also more nucleated. They have from thirty-one to sixty-seven surface rooms, with a mean of fifty-four rooms per site. Rooms are organized into linear unit pueblos or room blocks ranging in size from one to twenty rooms. Most room blocks had ten rooms or fewer. Rooms are relatively small and much more standardized during this period (Marshall and Walt 1984). They measure about four to nine square meters in size, similar to Salinas sites. The room block structures are spaced much more closely together than in the Tajo Phase, about ten meters apart. Like the Salinas jacals, these linear unit pueblos (room blocks) are usually one room wide, and occur in the familiar range of shapes: I, C, and F. Unlike in the Salinas area, however, there seems to be no preferred orientation for the room blocks.

There is no excavated information about the jacal sites in the Rio Abajo or from Salinas; this information comes from survey. If we assume, however, that each jacal room block includes a mix of habitation and also storage rooms, then one can interpret the typical small jacal site of less than a dozen rooms as representing an extended family unit consisting of several smaller groups, such as nuclear families (Kintigh 1994:132). According to this logic, most of the smaller jacal sites along the Rio Abajo were likely home for small numbers (one, two, or three) of family groups. However, the variation in room sizes during the Tajo Phase indicates considerable social variation in the size of the residential social group.

In contrast, the larger Early Elmendorf sites show two changes in social group organization. The smaller, more standardized rooms represent one unit of society, while the room block organization points to development of larger cooperative groups. Marshall and Walt (1984) note that the number of rooms and the total roofed area is about the same between the Tajo Phase and Early Elmendorf Phase. Therefore, it seems unlikely that this architectural change is tied to local population increase; rather, these observed architectural changes seem to indicate a change in social organization that led to a more standardized room size and more nucleated community structure.

Marshall and Walt (1984) interpret the unit structures (the contiguous room blocks) as representing an extended family group such as a lineage. If this inference is true, the growth in size of these room blocks by accretion seems to represent a change in the role of the residential group in village organization. In this case, newcomers would arrive in family groups of variable size and join up with established members of the same lineage. In this model, the growth of individual room blocks by accretion thus records incremental growth in the lineage group, primarily through immigration. In this model, lineage membership represents a significant social identity, affecting relationships between not only different kinship groups but also the established residents and the newcomers.

Late Elmendorf Phase Sites

The Late Elmendorf Phase here is dated to Pueblo III, about A.D. 1100 to 1300 (Marshall and Walt 1984:95). This time marks another major population shift in the Rio Abajo area. Most of the locales occupied by Early Elmendorf sites were abandoned, and people built large fortified masonry "apartment complexes" (pueblos), all in elevated and potentially defensive positions on the east side of the river. Only a few small hamlets remained occupied in more open areas. Marshall and Walt report that up to 95 percent of the population crowded into these large masonry pueblos. This population concentration represents a highly nucleated settlement strategy.

This time period spans the transition between the Jacal period and the Early Pueblo period in Salinas. The details of the Salinas pueblos are discussed in the next chapters, but it is important to note here that the early pueblos in the Rio Abajo exhibit much more variation in their form and layout. Rooms vary in size even within a single site. Rooms of smaller size (two to four square meters) typically are located adjacent to open courtyard areas, but the rooms abutting them may be as large as twelve or fifteen square meters. At some sites, there are differences in construction as well. Shaped masonry construction was used to form the rear-facing walls, but walls that faced the courtyard were typically built of clast-based jacal, made with rounded river cobbles. Late Elmendorf Phase sites also show a loosely defined central courtyard or plaza. Marshall and Walt (1984:97) interpret this spatial organization as representing "incipient development of plaza organization."

Marshall and Walt (1984) suggest that the difference in room size, location, and construction methods may indicate a difference in room function. If the larger rooms are living structures, the smaller, cobbled-walled ones adjacent to this open courtyard may have been used for storage, or for animal pens. In the Southwest, animal pens usually refer to housing for domesticated turkeys (e.g., Lang and Harris 1984; Olsen 1990), which were sometimes kept in small masonry pens, but we do not know if that practice was common in more ancient times.

It seems clear, however, that there is a real increase in population from the Early Elmendorf Phase to the Late Elmendorf Phase. Marshall and Walt (1984) report a dramatic increase in both the number of rooms and in the total roofed area per phase. There are a total of 252 rooms on sites dating to the Early Elmendorf Phase, and 443 rooms associated with Late Elmendorf Phase sites—a 76 percent increase in rooms and a 98 percent increase in total roofed area. Marshall and Walt (1984:97) interpret this magnitude of difference as evidence of significant population immigration. They caution, however, that there is no obvious sign of imported objects or ideas that these presumed newcomers would have brought with them when they arrived. In addition, one should note that the Late Elmendorf Phase represents a longer period of time, and many sites span both Early and Late phases. So it is not altogether clear what scale of immigration is represented here. At the same time, the high degree of diversity in site organization in Late Elmendorf sites does indicate that people exhibited much less shared knowledge or expectations regarding site plan during this time. The villages during the Early Elmendorf Phase were much more similar to one another, suggesting a greater degree of mutual shared knowledge among the groups living in the region during the earlier time period. The higher diversity observed for Late Elmendorf Phase sites could indicate that later immigrants adhered less closely to these established norms (Marshall and Walt 1984:98). Marshall and Walt (1984:98) comment that this apparent lack of standardization in Late Elmendorf Phase site organization contrasts with the "near formulaic character" observed in the Early Elmendorf Phase villages during Pueblo II and the large late pueblos of Pueblo IV.

At the time when in the Salinas region people were building early masonry pueblos and importing glaze ware ceramics, the Rio Abajo experienced a second large population increase, with an estimated sevenfold increase in population from about A.D. 1300 to 1400 (Marshall and Walt 1984). In addition to whatever immigrants moved in from outside the area, the local inhabitants also moved back down into the river valley and built large plaza-oriented pueblos of puddled adobe. In the fifteenth century some of these large multistoried terrace pueblos had two hundred to six thousand rooms and housed up to two thousand people. In the Rio Abajo, this period from the introduction of Rio Grande glaze ware pottery after A.D. 1300 to the time of Spanish contact soon after 1540 is called "Ancestral Piro" (Marshall and Walt 1984:135–136).

Where did these proposed immigrants to the Rio Abajo come from? Marshall and Walt (1984) point out that after about A.D. 1100 there are a number of population changes in the Southwest, with regional abandonment of areas around Chaco Canyon and Mesa Verde (see Cordell et al. 2007 for a more recent perspective). We do not know whether people from the northern Southwest actually moved into the Rio Abajo; the point is that widespread regional population movements apparently affected settlement along the Rio Abajo. The impact of this process on the Salinas area, however, is still unknown.

Comparing Rio Abajo and Salinas Jacal Villages

Early village sites in both the Rio Abajo and Salinas regions show a change from pithouse architecture to basic pueblo construction, including adoption of masonry construction, aggregation of small structures of one or two rooms into linear contiguous room blocks, and greater formal arrangement of room blocks in relationship to a communal space or plaza. In contrast to the Salinas pattern of long-term stability of settlement within a given locale, however, the Rio Abajo region saw considerable population movement, both within the region and also in the form of significant immigration.

On the scales of the village sites themselves, there is a comparable shift in room size in both regions, from larger structures in the earliest periods to smaller (about four square meters) room sizes in later periods. In many other respects, however, Salinas villages seem to record a narrower range of variation in several other socially sensitive variables, including room size, room block construction, village site spatial organization, and also village site nucleation.

Comparing the evidence from the built environment in both these regions highlights the extent to which the Salinas area may be fairly unusual in the degree of standardization or homogeneity in these aspects of the built environment. By the time people were building jacal surface structures in Salinas, they shared some specific concepts of village organization, not just within a site but also within the region as a whole. As an example, room size in the Rio Abajo seems to be quite variable within sites, within time periods, and even across time periods. This variation may be associated with room function. However, we might expect that Salinas farmers would have had comparable needs for different room functions for living, storage, and animal pens. Yet these different needs or functions do not translate into the same range of variation in room sizes in Salinas. Compared to the Salinas area, these observed sites along the Rio Abajo appear to be much more temporary settlements, constructed with more than one set of assumptions about community organization, and with these culturally assumed rules for social life apparently modified readily over time as needed.

These village sites present a very different concept of social organization within the village community. The variable organization of village communities in the Rio Abajo may express people's rather ad hoc adjustments to continuing changes in the local social environment in the context of these repeated population movements both within the region and from outside. While a high degree of variation in the built environment can indicate a lack of social integration at the group level, and hence greater fragility of the group as a whole, this apparent flexibility and adaptability could also have been a positive influence, allowing the people along the Rio Abajo to cope with fairly rapid and repeated demographic changes.

Jacal Village Society in Salinas

This brief look at the neighboring region to the west provides a surprisingly different social strategy for early village community organization. In the Salinas region, there is evidence of continued occupation in specific locales for many hundreds of years. The spatial organization of jacal structures in the uplands show that the people who lived on each site had a number of shared ideas about their social organization, group size, and the use of space.

It is also apparent that the Salinas population as a whole shared certain views and a common experience of occupation within the region. Each village site shows continuity of occupation in a given locale, a continuity that persists into the Early Pueblo period. There is apparently a region-wide and long-standing preference for settlement in elevated situations. These elevated settings include isolated ridges, such as Turkey Ridge, or locales along the rims of the mesa itself, as in the jacal component of LA 9032, or along the rims of local promontories in the interior of Chupadera Mesa, as seen for LA 197. Widely shared ideas about general size, shape, and organization of rooms, room blocks and also room block groupings contribute to the sense of a shared history and experience of all groups within this region.

Chamberlin (2008) also reports comparatively little evidence of the kind of social activities that would highlight differences between the several village communities. For example, in the Salinas area, we do not see so much effort spent in obvious territorial boundary marking with rock art, shrines, or cairns. I do not mean to imply that there is no such evidence in Salinas; in fact, Abó Pass is well-known as a rock art locale, with many different and extensive panels of painted and pecked images (Cole 1984; Schaafsma 1993). There are also other panels of rock art near Pueblo Blanco and in Arroyo Seco. Chamberlin (2008) also notes the presence of shrines. However, it is also true that the density of territorial marking seems low compared to other areas of the Southwest (compare to Duff 2004; Fowles 2009). Relatively continuous and unrestricted communication of ideas as well as artifacts and raw materials are also apparent in the limited array of design-style attributes of Chupadero Black-on-White pottery across the region (Clark 2006; Chamberlin 2008).

The diversity in size and organization between different jacal sites, however, shows that this overall regional uniformity accommodated local variation in the exact size, composition, and spatial organization of the village site. At one spatial scale, Caperton (1981) notes a distinction between the smaller northern jacal villages by Abó and the larger southern villages by Gran Quivira. Within the southern group itself, Chamberlin (2008) also notes a range of variation in the size of rooms, the number of rooms in the largest room blocks within a site, the degree of nucleation of room block structures, and also the organization of structures in relationship to communal open spaces (Chamberlin 2008:93). Again, these architectural features represent locally variable aspects of village organization within the overall regionally shared values that gave primacy to, for example, a north-south orientation, and additions of "extra" lines of rooms oriented to the east. This local differentiation at the level of the site seems to point to a regional social network of continuing interaction, but one in which the different village communities represented fairly well-defined, bounded, and coherent social groups whose members shared a history of co-residence and a sense of place within a local landscape.

Within a village community, the architecture of room blocks follows a similar idea of diversity within a given range of variation. Internal site variation in living arrangement is expressed in the diversity of sizes of room blocks, their shape, and spatial relationship to one another. On the other hand, Chamberlin's (2008) maps of the various sites show that there is no apparent patterning in the distribution of room blocks of different size—for example, there is no indication that room blocks of a certain size cluster together, near the open (plaza) spaces, or near the edges of the settled area. These small pieces of evidence suggest that local residential groups or families controlled their own space allocation and building within the village site. Local-level decision-making is also seen in the record of ceramic exchange (Chamberlin 2008). Each jacal site yielded decorated pottery derived from several different sources, for both local and non-local pottery types.

In general, however, the most abundant sources for local pottery were those that were closest to each village. This evidence indicates that distance and transport costs may have been the primary factors affecting people's choice of pottery source (Chamberlin 2008). In social terms, we might infer that people in each village community had their own preferred sources for obtaining resources, raw materials, and trade goods, but that relationship did not preclude access to different sources. In this respect, then, different subgroups within each jacal village community seemed to operate more or less autonomously, but within the boundaries of certain expectations and assumptions that were shared widely across the region.

However, the variety of ceramic source areas represented at all jacal sites does indicate a smaller-scale but continuing degree of interaction between the various village communities throughout the region. We do not know whether this interaction represents relatively informal individual or family group decision-making, or whether larger social groups were involved as more formally arranged social events. We know that during Pueblo IV, people at Gran Quivira and the other large pueblos hosted relatively formal trading events and interpueblo feasts (e.g., Spielmann 2004a). It is possible that the beginnings of communal spaces that we do see at some sites during the Jacal period may relate to such regional gatherings.

In sum, the available information from Salinas jacal sites fits well within a basic model of community organization that recognizes that people in each jacal village site exhibited aspects of community that included, among other things, some degree of self-identification and local autonomy as a local village community within a cultural landscape that included continued interaction with similarly organized but related local groups. Differences in the size of residential groups within a village site are obvious, with some room blocks being larger than others. In addition, the jacal sites on Turkey Ridge generally resemble one another more than they do the other jacal sites on Chupadera Mesa (Chamberlin 2008:93). However, compared with the much higher range of variation seen among sites of comparable time period in the Rio Abajo, the observed differences between the jacal communities in Salinas seem to be explained best as local variations within fairly consistent ideational boundaries set by shared cultural assumptions.

CHAPTER FOUR

Early Pueblos and Communities

The Early Pueblo period in the Salinas area dates from about A.D. 1150 to 1300, a period that corresponds to the Pueblo III period in the northern Southwest. As with the transition between pithouse and jacal architecture, however, there is apparently some temporal overlap in a chronological sequence that is defined primarily by architectural construction methods and site layout. Toulouse and Stephenson (1960) first attempted to make sense of the local chronology, distinguishing the lowland Claunch Focus jacal sites from the Arroyo Seco Focus sites in the uplands. Their basic temporal framework is shown in table 10, with comments included from Stuart and Gauthier (1981:table 7.5) and also my own research. In this scheme, the Early Pueblo period is represented by a temporal phase that they called the Gran Quivira Focus, a time period characterized by masonry pueblo construction and also the import of Rio Grande glaze wares.

Archaeological research since Stuart and Gauthier's (1981) summary assessment indicates that the transition from jacal structures to nucleated pueblos represents an even more complex process. Caperton (1981) introduced the terminology of "transitional structures" to describe a change in construction methods that he observed on some late Jacal period sites; these jacal structures showed greater use of masonry construction. In addition, my excavations at Kite Pueblo (LA 199) in 1994 revealed yet another type of site in the Salinas region: the plaza-oriented adobe pueblo.

In this chapter, I consider the evidence of community organization during this time of transition between the Jacal period and the construction of masonry pueblos. Our best understanding of this period comes from my excavations at Kite Pueblo (LA 199). Until recently, Kite Pueblo was the only known adobe pueblo in Salinas—until my later excavations at Frank's Pueblo (LA 9032) unexpectedly revealed architectural evidence of a very similar adobe village under later masonry construction. Figure 13 shows the location of these two known early pueblo sites; the other two sites shown in this figure (the "Black-on-White pueblos") will be discussed in more detail in the next chapter.

In the Salinas Province, one interesting local puzzle is why people would build adobe pueblos in the first place. While adobe pueblos are well-known from sites along the Rio Grande, including, for example, the pueblo at Coronado State Monument north of Albuquerque, one might think that the Salinas residents, lacking a ready supply of water, would have preferred to use local stone. In fact, stone suitable for construction is much more readily available than is water for making adobe. In the northern portion of the Salinas region, the red sandstone of the Abó formation outcrops in tabular slabs, while upland areas of Chupadera Mesa include many exposures of limestone from the San Andres formation (Bates et al. 1947). On the other hand, we can also consider adobe architecture a logical extension of the wattle-and-daub construction used in jacal structures, with a reduction in the wattle (sticks) and an increase in the use of daub (adobe).

However, it is apparent that these architectural choices involved other consideration. I hope that it is clear that in the Salinas region, there is no single or simple progression of construction technique from jacal to adobe to masonry structures. In fact, the jacal sites that are latest in time are dominated by masonry, not adobe. These so-called transitional structures represent an altogether different kind of combination of building techniques; in this case, the upright wattle sticks acted as framing studs for stacks of stone blocks (not adobe) in between. Similarly, the adobe pueblo can hardly be considered a stage in people's development of the concept of a masonry pueblo, but rather represents a constructional strategy in its own right. As I hope to show in this chapter, the adobe pueblo also represents a social strategy—one emphasizing the pueblo community as a social unit. I suggest that this idea of the adobe pueblo represents a broadly shared vision of community organization that is expressed in architectural form.

Kite Pueblo (LA 199)

Kite Pueblo, a compact adobe plaza-oriented pueblo, did not have a confusing overlay of extensive masonry architecture from later occupation. It

Table 10. Phase sequence of the Gran Quivira subregion

Claunch Focus	
Pecos time period	Pueblo III, about A.D. 1200 to 1300
Typical ceramic types represented	Jornada Brown; Los Lunas Smudged; San Francisco Red; Chupadero Black-on-White; Indented Corrugated Utility (Stuart and Gauthier 1981:table VII.5)
Comments	Stuart and Gauthier (1981) describe these sites as small, with one- to three-room surface structures and no definite plaza. They emphasize that St. John's Polychrome is noticeably absent. Toulouse and Stephenson (1960:40) note that Claunch Focus sites are at lower elevations; Stuart and Gauthier (1981:323) suggest that they date to slightly earlier in time. They believe that Claunch Focus sites represent greater affinity with southern (Jornada Mogollon) groups (Stuart and Gauthier 1981:323).
Arroyo Seco Focus	
Pecos time period	Pueblo III, about A.D. 1200 to 1300 (contemporary with the Claunch Focus)
Typical ceramic types represented	Chupadero Black-on-White; St. John's Polychrome; Corrugated Utility (Stuart and Gauthier 1981:table VII.5)
Comments	Toulouse and Stephenson (1960) suggested that Arroyo Seco Focus sites were contemporaneous with those of the Claunch Focus, although located at higher elevations. Stuart and Gauthier (1981:323) believe that the presence of St. John's Polychrome on these sites argues for a slightly later date. They also suggest that these sites differ from those of the Claunch Focus by showing greater affinity with northern (Anasazi) Pueblo Indian groups (Stuart and Gauthier 1981:323).
Gran Quivira Focus	
Pecos time period	Early Pueblo IV, about A.D. 1300 to 1425
Typical ceramic types represented	Chupadero Black-on-White; Agua Fria Glaze-on-Red; San Clemente Glaze-on-Polychrome; Cieneguilla Glaze-on-Yellow; Jornada Brown; Indented Blind Corrugated (Stuart and Gauthier 1981:table VII.5)
Comments	Stuart and Gauthier (1981) note that this time period represents an influx of influences from the Jornada Mogollon (as seen in Jornada Brown ware) and also from the north (Rio Grande Glaze A ceramics). They note a change in site location to "easily defensible positions such as mesa tops or ridge crests" (Stuart and Gauthier 1981:323–324). These sites represent the "Glaze A pueblos" and include the late masonry component of Kite Pueblo (LA 199) and also at Frank's Pueblo (LA 9032). The main occupation at Pueblo de la Mesa (LA 2091; chapter 6 in this volume) is also considered a Glaze A pueblo.

The original phase names were defined by Toulouse and Stephenson (1960), based on fieldwork associated with their excavations at Pueblo Pardo in 1941. Stuart and Gauthier (1981:table VII.5 and pp. 331–324) added the dates and time period designations.

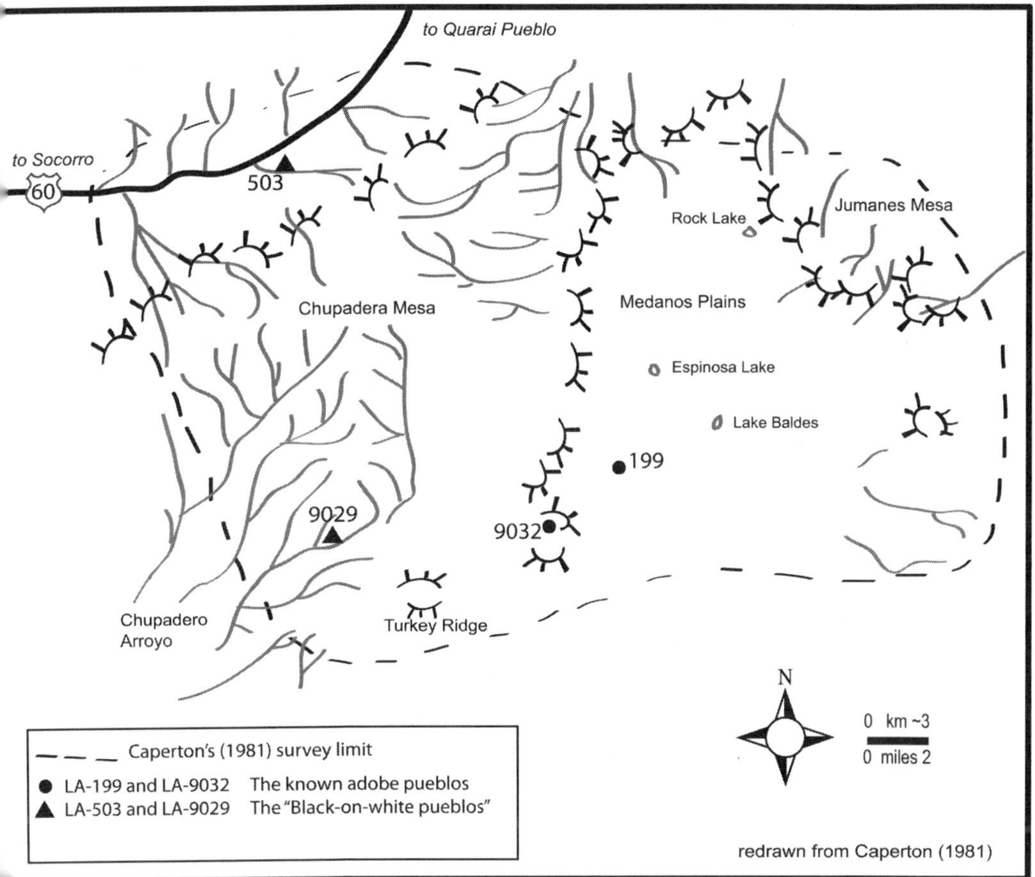

Figure 13. Location of early pueblo sites in the Salinas Province. Only the pre-Glaze A pueblos are shown. Modified from Caperton (1981:xii).

was therefore particularly useful in investigating the role of this standardized site layout during regional migration in the twelfth and thirteenth centuries. Here, an investigation of the history of the architecture of the site tells us not only how it was built, but it also reveals the social and cultural context within which people made decisions about creating, using, and modifying their built environment.

Kite Pueblo is located in a pasture about a kilometer away from the pueblo of Gran Quivira, and within two hundred meters of Kite Pithouse Village. Both archaeological sites on the Kite family ranch are situated in the sandy Medanos Plains, just east of Chupadera Mesa. The pueblo is a compact single-story structure, organized in four room blocks surrounding a central plaza. Figure 14 shows the overall site layout, the terminology that we used for different parts of the site, and the locations of our excavation units. The details of excavation and analysis are reported in three descriptive reports (Rautman 1995a, 1995b; Rautman and Heyman 2011), and also in Rautman (2000).

The vast majority of ceramic sherds present at the site were identified as Corona Corrugated brown ware and also Chupadero Black-on-White. As discussed in chapter 3, Chupadero Black-on-White is a widespread ceramic type that is found throughout central and southeastern New Mexico; its long production period dates from between about A.D. 1150 and 1550 (Wiseman 1986; Clark 2006). Because of this long production period, the simple presence of Chupadero Black-on-White does not provide precise chronological information. Rather, in the Salinas area, one looks for the presence or absence of the well-dated imported Rio Grande glaze ware types to establish site chronology (Hayes et al. 1981:90–103).

At Kite Pueblo, only a very few Glaze A sherds were found; they were all Agua Fria Glaze-on-Red, which was produced in the early 1300s (Warren 1980; Hayes et al. 1981:93; Eckert 2006). These sherds were also concentrated in loose surface sediments or the uppermost portions of room fill. It seems clear that use of this pottery coincides with the latest occupation of the pueblo. No Glaze B pottery was found. This association suggests that the pueblo was abandoned early in the fourteenth century, before the production of Glaze B pottery about A.D. 1410 (table 14 in the next chapter provides a chronology of the glaze ware pottery types). During this relatively short span of time—about 150 years or so—a group of people built and lived at Kite Pueblo.

Radiocarbon dates from the site offer no particular surprise: they are consistent with Caperton's (1981) estimates and the evidence from the surface ceramic assemblage (table 11).

Evidence of Daily Life at Kite Pueblo

Analysis of the excavated artifacts and paleobotanical remains and animal bone from midden contexts and room fill also was fairly unremarkable, confirming (as one would expect) that the people at Kite Pueblo were farming domesticated maize, beans, and pumpkin/squash. The pollen evidence from the archaeological deposits indicates that the piñon-juniper forest of today was considerably more open at the time of site occupation (Cummings et al. 1997). This change in the density of the forest cover

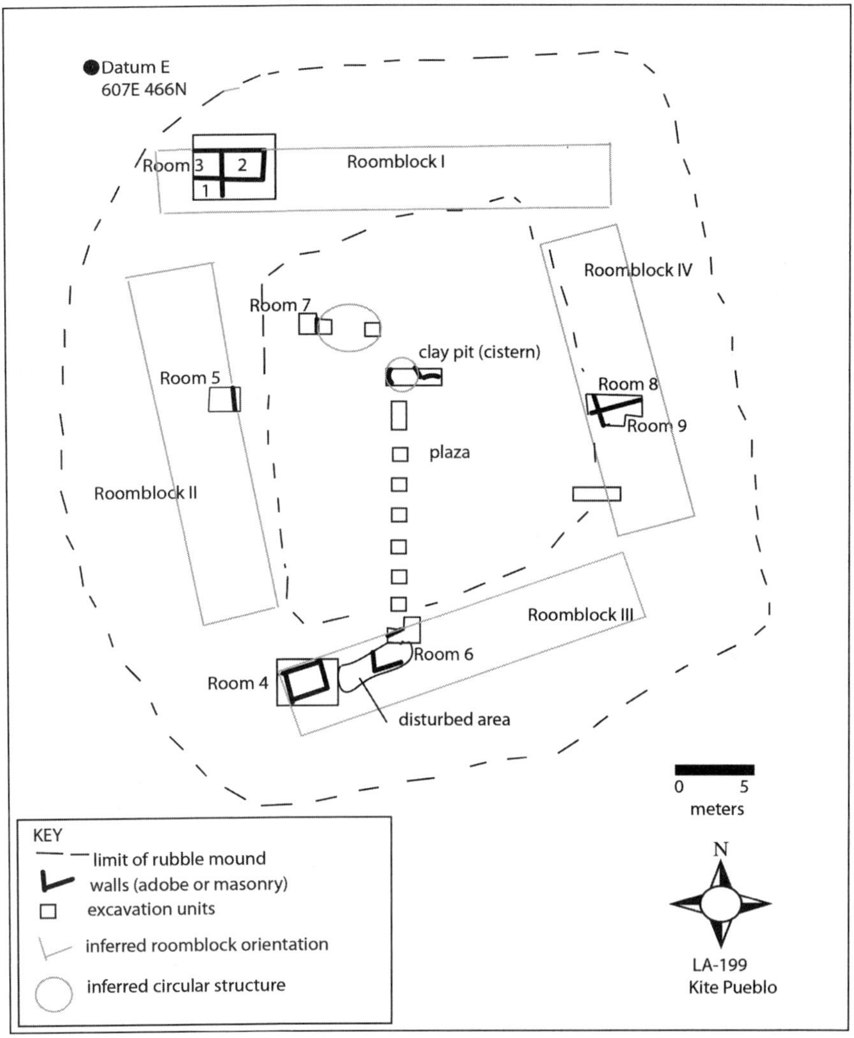

Figure 14. Map of Kite Pueblo, LA 199. The room blocks are two rooms wide, but their exact lengths are estimated.

is likely due to human activities around the pueblo itself, including use of wood for construction and fuel. The plant remains from the site show that, like earlier pithouse dwellers, the people who lived in Kite Pueblo also utilized a wide variety of wild plants for food and other purposes. The most ubiquitous edible plant remains include different species of cactus and several types of seed-bearing annuals. Animal taxa include mule deer,

Table 11. Radiocarbon dates, Kite Pueblo, LA 199

Beta Number	*Context*	*Uncalibrated Age B.P. (± 1 σ)*	*Calibrated Date A.D. (± 2 σ)*	*Intercept A.D.*	*Intercept Date cal. A.D. (± 1 σ)*
94153	North Room block, Room 2; seed; below the hearth; before early occupation	1120 ± 50 B.P.	A.D. 785 to 1020	A.D. 960	A.D. 880 to 995
94152	North Room block, Room 2 hearth; charcoal; early occupation	840 ± 60 B.P.	A.D. 1035 to 1285	A.D. 1220	A.D. 1170 to 1265
94151	North Room block, Room 2; charcoal; later occupation	730 ± 50 B.P.	A.D. 1225 to 1310; A.D.1355 to 1385	A.D. 1285	A.D. 1265 to 1295
94150	North Room block, Room 2; maize; later occupation	740 ± 50 B.P.	A.D. 1220 to 1310; A.D. 1365 to 1375	A.D. 1280	A.D. 1260 to 1295
94154	South Room block; Room 4; maize; just above floor of remodeled adobe room	660 ± 50 B.P.	A.D. 1275 to 1410	1300	A.D. 1290 to 1325; A.D. 1340 to 1390
94156	East Room block, Room 8; maize; just above floor	680 ± 50 B.P.	A.D. 1265 to 1400	A.D. 1295	A.D. 1285 to 1310; A.D. 1355 to 1385
94155	Plaza structure, charred twigs; just above floor	680 ± 60 B.P.	A.D. 1250 to 1410	A.D. 1295	A.D. 1280 to 1315; A.D. 1345 to 1390
94158	Clay pit in plaza, charcoal; near the base	690 ± 100 B.P.	A.D. 1175 to 1430	A.D. 1295	A.D. 1260 to 1400
96094*	Clay pit in plaza, charcoal; near the top	1040 ± 70 B.P.	A.D. 880 to 1170	A.D. 1005	A.D. 970 to 1035

*Sample 96094 at the top of midden within the clay pit is the only one that does not fit into the observed stratigraphic sequence; it may be that the later occupants of the site deliberately filled the clay pit. If the topmost sediment in the clay pit was obtained from another area of the site, it could well include cultural debris from the earliest occupation of the site. (Radiocarbon dating by Beta Analytic, Inc. Radiocarbon Dating Service, Miami, Florida.)

pronghorn, and also smaller animals such as cottontail rabbits and jackrabbits. A small number of bison bones were also recovered (Rautman 1995a; Rautman and Heyman 2011).

All of these animals are available near the site, with the possible exception of mule deer and bison. Mule deer are associated with more upland forested area, but do sometimes browse at lower elevations. Bison (buffalo) are generally found on the open plains farther east, although historic records note that occasionally small groups or individual buffalo would roam this far west (Tom Carroll, National Park Service superintendent, personal communication 1986). Bison remains on Salinas sites are thus considered to come from some distance away, through trade or long-distance hunting treks (e.g., Hayes 1981; Spielmann 1996; see also Speth and Rautman 2004).

Altogether, the plant and animal remains at both Kite Pithouse Village and nearby Kite Pueblo show a great deal of continuity in many aspects of life across more than two hundred years. Despite this continuity in basic economic strategies, significant changes in social and political organization were occurring at the time Kite Pueblo was constructed. Observation of the constructional history of the pueblo and the details of the built environment allow a much more detailed look at the social processes and shared ideas that structured community life during this period.

The Construction of Kite Pueblo

The history of the pueblo is important here in establishing the physical parameters of the built environment for the pueblo's occupants, which in turn provides considerable information about social dynamics over the period of site occupation. At this site, we can trace the historical and social process involved in the physical construction of this village community over time. Because we can identify the sequence of decision-making in some detail, it is possible to examine specifically how people used and modified their built environment in the context of changing social strategies within their community.

The architectural details at Kite Pueblo reveal at least two constructional phases: an early phase characterized by use of adobe, and a later phase of masonry construction using limestone (table 12). The first adobe pueblo consisted of four sets of contiguous rooms (room blocks), two rooms across and about eight to fifteen rooms long, for a total estimate of roughly sixty to one hundred rooms. The walls were made of adobe blocks irregular in size, but typically measuring no more than about thirty centimeters

Table 12. The history of Kite Pueblo, LA 199

Sequence	*Evidence/Location*	*Interpretation*
1. Pithouse occupation	Curved caliche wall below the north room block	Pit structure wall (?) below an adobe room; this structure may be part of LA 38448, Kite Pithouse Village
2. Adobe pueblo	Adobe walls found in Room 2 and in all room blocks	Adobe pueblo; evidence of adobe rooms in all four room blocks around a central plaza; adobe room size estimated at 2.3 m E-W by about 2 m N-S
	Clay pit in plaza	Cistern used during occupation; then filled with midden sediments, showing continued occupation of the site
3. Masonry remodeling	Masonry walls built atop adobe walls; observed on all four room blocks	Evidence of masonry wall footings on top of the old adobe walls; most are extensively eroded; visible as rubble on the ground surface
	Rooms added to east side of the pueblo	Second strategy of remodeling: adding rooms to the masonry pueblo by building new masonry rooms on the exterior of the room block
	Masonry pit room(s) located in the NW corner of the plaza	Construction of at least one semi-subterranean square room; well-made masonry wall distinguishes it from the Room block IV construction.

long and fifteen centimeters thick (top to bottom). These blocks are likely comparable to the handmade "tablets" or "elongate clumps" of adobe described by Judd (1916:242–244) at other prehistoric sites. Well-preserved level adobe floors of about two centimeters thick lapped up onto these walls.

We were able to excavate about half of one undisturbed early room. In other parts of the pueblo, most of our efforts were directed toward clearing loose sand from areas of modern disturbance (e.g., backhoe scars) in order to identify the presence of adobe walls and their spatial relationships. In the one room that we tested on the north side of the pueblo, we found evidence of a curved earthen wall about a meter below the ground surface. This curved portion of wall may represent a pit structure that was dug into the subsurface caliche and thus predates construction of the pueblo. Its presence is not particularly surprising given the proximity of Kite Pithouse Village in this same pasture. Alternatively, field observations note that it could also be a section of slumped adobe wall or highly eroded roof fall from an adobe surface structure. At any rate, excavation in this area also identified three clearly defined adobe walls and a well-preserved floor with a neat round hearth, hemispherical (bowl-shaped) in cross-section. The depression was dug down into the floor and then plastered seamlessly with the floor adobe. It was filled with fine white ash; radiocarbon dates from this material place it as expected within the earliest occupation of the pueblo, closer to A.D. 1150.

Our excavation unit did not extend to the southern wall of this adobe room. Assuming that the hearth was likely situated near the middle of the room, we can infer that these early rooms were likely rectangular in shape. Measuring from the plaza to the pueblo perimeter (that is, north-south across the room block), the rooms would be about 2.3 meters in length (parallel to the length of the room block) and about 3.5 to 4 meters across.

Similar adobe construction on all four sides of the pueblo shows that the pueblo was built as a unit, as a preplanned and organized single event. This was decidedly not a pueblo that was constructed by accretion over time. Rather, the spatial relationships of the walls, the similarity of construction (adobe), and the bonding and abutting patterns visible in the portions of walls that we exposed in all four room blocks show that construction adhered to an overall plan. A suite of early radiocarbon dates indicates that this early aboveground construction began just before the thirteenth century.

A later remodeling period associated with this adobe pueblo involved construction that added rooms while preserving the same overall site layout. There was also some effort made to preserve and even expand the useable open space in the central plaza. The south room block shows the clearest evidence of this effort, where one adobe-walled room, and presumably the entire room block, was shifted about one meter to the south, away from the plaza. The northern plaza-side portion of the room was removed. The south perimeter adobe wall of the earlier structure was removed, and the stubs of the adobe wall on either end of the room were incorporated into the north-south walls of the new room. Two new east-west walls were constructed to form a new room, and a new floor was built up of thick adobe in the south half of the room to match the height of the older floor. The walls were then replastered with adobe, which covered the visible differences of adobe block construction of the two building episodes.

This remodeling represents a great deal of energy and effort spent in razing and constructing walls, with comparatively little gain in open plaza space. For this reason, one might wonder if the observed remodeling was not itself the primary goal of construction, but rather a consequence of some other decision. Whatever the reason or goal, however, the end result was in fact an increase in the length of the plaza from north to south.

An increase in the east-west width of the whole pueblo is also shown on the east side, where additional rooms were added onto the outside of the pueblo, away from the plaza. These additional rooms consist of masonry and also adobe walls that extend east from and abut onto the earlier, adobe-walled rooms. It is not clear whether the earlier and later rooms were occupied at the same time, or if the older rooms were then abandoned or razed to increase the plaza space. It is also not clear whether the new rooms had entryways leading outside the pueblo or whether they were accessible only through the plaza.

In contrast to the original construction of the adobe pueblo, this remodeling in the east room block was not at all standardized in construction method. In fact, each wall incorporated a different proportion of masonry and adobe. This variation indicates that the different walls were not built on as a unit, but could have been constructed at different times or by different groups of people with different concepts or resources for room construction. The effect of this remodeling, however, was an increase in the potential habitable space in this room block while retaining the open space of the central plaza.

A different kind of later remodeling may be more accurately considered rebuilding. In this second construction episode, the early adobe rooms had either been abandoned and then reoccupied, or at least repurposed. These later rooms were basically square or slightly rectangular in shape, about two meters in each dimension, and made with masonry block construction. They can be seen in the north room block (Room 2) and in the south room block (Room 4). Scattered blocks of masonry rubble on the ground surface indicate that similar rooms were likely built on the west room block as well.

In Room 2, the masonry construction appears to post-date occupation of the underlying adobe room. Here, a well-preserved adobe floor is associated with, and laps up onto, the masonry walls. The articulated remains of a turkey were found resting on the floor of the room (0.6 meters below the ground surface). In contrast, in Room 4, the masonry walls were built directly atop the older adobe wall on three sides (north, east, and west) of the earlier adobe room, resulting in a room where the lower portions of the walls were adobe and the upper portions were masonry. In this room, the only clear floor is that associated with adobe walls. In this case, it appears that the original adobe room was remodeled to include masonry in the higher portions of each wall.

Most of the masonry of the late phase walls in Room 4 had eroded and the walls disaggregated over time, leaving only two or three courses intact, where they were covered and protected by windblown sand. The amount of rubble on the ground surface indicates, however, that this remodeled room with partial masonry walls was likely no more than one story high. It is important to realize that the pueblo is located on the sandy lowlands of the Medanos Plains, and limestone is not available in the immediate site area. The closest source for rock would be about a kilometer or so away, where a rocky ridge extends out from Chupadera Mesa—not exactly far, but still, every block would have to be carried in. Gran Quivira Pueblo sits atop this ridge, but we do not know if there was a comparable adobe pueblo at Gran Quivira during this time period or whether there were any social restrictions on people's access to limestone sources.

On the south side of the room, the masonry wall was constructed over earlier cultural fill, indicating that the masonry room was smaller in size than the previous adobe one, similar to the situation seen in Room 2 on the north room block. Then the masonry room was finished with a well-made adobe floor (now highly eroded) that had lapped up onto the wall blocks. Most of the Glaze A sherds that were found at the site came from the ground surface; some also were found in the upper levels of room fill in this area.

Similar masonry walls occur atop the adobe rooms on at least three sides of the pueblo: north, south, and west room blocks. This kind of building or rebuilding "upwards" is not recorded in the east room block. The surface rubble in this area is also visibly less in this area, suggesting that in fact this late masonry rebuilding was confined to the three other room blocks.

In sum, the constructional history of the pueblo shows at least two periods of occupation. In the first phase, an early plaza-oriented adobe pueblo was built with four room blocks surrounding a central square or rectangular plaza. People did some remodeling to the basic pueblo, including adding rooms on the east side of the pueblo and shifting the location of rooms on the south. The second occupational phase is characterized by masonry construction building up from established adobe walls. Some rooms reused the floors of the older adobe occupation, but some later rooms involved new floor construction, above the level of the older adobe ones. These later rooms are smaller in size and square rather than rectangular in shape.

The association of glaze sherds with the uppermost levels of sediment at the site suggests that masonry construction likely occurred just before or just after the first import of Glaze A pottery into Salinas. The relatively small number of Glaze A sherds for the site as a whole, however, indicates that the site was also abandoned soon after this time.

It is possible that the pueblo was abandoned for a short time and that the masonry rebuilding represents a reoccupation, rather than simple remodeling, by the then-current residents. However, the construction of three of the four walls of Room 4 directly atop the underlying adobe walls of the earlier structure argues for a fairly high degree of continuity in site use, or a fairly short period in which any exposed adobe would be subject to erosion.

Exterior Space and Plaza Activities at Kite Pueblo

The four room blocks at Kite Pueblo outlined the perimeter of a rectangular central plaza. One important discovery was that the central plaza was not in fact an empty exterior space; rather, two subterranean (or semisubterranean) structures were located in the plaza, at least during the initial period of site occupation.

In some areas of the northern Southwest, a pit feature in the middle of the plaza would be readily interpreted as a kiva or other ceremonial structure. This pit structure at Kite Pueblo, however, is most accurately identified as a cistern or water storage feature, such as those described for Gran Quivira (Howard 1981). It had a pear-shaped (rather than circular) outline, with the small end (on the east side) forming a caliche step down into the pit. The large end of the pit measured about 1.5 meters deep and was lined with 2 to 5 centimeters of fine clay. An adobe wall had been built at the rim of the small end of the feature, which would have extended above the plaza surface. This adobe wall may have encircled the feature and provided additional area for accumulation of midden after the pit itself had been filled.

Experiments at Gran Quivira, reported by Hayes et al (1981:20–22), confirm that such clay-lined pits could in fact retain water. It seems likely that this possible cistern at Kite Pueblo was constructed and used during the early occupation of the site, and that it ceased to be used (and filled with cultural debris) later in the occupation of the site, but before the masonry rebuilding period.

A second subterranean or semisubterranean structure was also identified in the plaza (Room 7). Here we found a deeply buried adobe floor two meters below the present ground surface, with about 1.5 meters of wall preserved above. A second test pit to the east uncovered apparent roof fall. No roof beams were preserved, but charred twigs within cultural fill date to the later occupation of the site. A few undecorated red ware sherds were found inside this structure, but none with evidence of glazed decoration. This structure might represent one large room, or else possibly a series of smaller roofed rooms spanning across the plaza, adjacent to the north room block. The information available cannot distinguish between these options, but in either case, this masonry construction is the only evidence of construction within the plaza itself. Both the cistern and the later room (or rooms) in the plaza fell out of use and were filled with midden and aeolian sediments before site abandonment.

In sum, the architecture of the site, together with the chronometric information from pottery and radiocarbon dates, indicate that the plaza-oriented layout of the site was important from the earliest period of occupation and was preserved throughout the occupation of this pueblo. This continuity of architectural layout suggests that the central space of the plaza played an important role in this community.

For many years, Kite Pueblo was the only known adobe pueblo site in Salinas where we had any excavated information. Caperton (1981) reports that Axtell Pueblo and possibly Lost Pueblo date to roughly the same span of time (before the import of Glaze A ceramics), but we have very limited information about these sites (see table 13). Recent excavations at Frank's Pueblo (LA 9032), however, unexpectedly revealed the presence of a plaza-oriented adobe pueblo of similar size underlying the surface masonry architecture. The small amount of information we have about this buried adobe pueblo is described in the following section, while more details of our excavations at Frank's Pueblo are presented in chapter 6.

The Adobe Pueblo at LA 9032 (Frank's Pueblo)

For each of my excavation projects reported in this volume, the research that I intended to pursue at a given site was not necessarily feasible with the data that I found. My excavations at Frank's Pueblo were possibly the most closely tailored to answer questions of sociopolitical organization of village life, yet the site ended up being so different from what was expected that once again the excavations presented far more new questions than answers.

In this case, I was investigating evidence for corporate group activities, particularly in artifact production, distribution, and use. Based on my previous excavation experience in the region, I anticipated finding extensive trash deposits in different portions of the site, each yielding abundant quantities of many different types of artifacts. At Frank's Pueblo, however, few artifacts were found in expected contexts such as room fill and midden. Moreover, most of the midden that we did identify was located on sloping ground where the ceramic and lithic artifacts, and animal bone, had likely experienced extensive downslope erosion and weathering. The comparability of this midden information to that contained within thick in situ midden deposits of other sites is therefore highly debatable.

The architecture of the site, however, proved to be of considerable interest. Chamberlin's (2008) survey had mapped a dispersed Jacal period occupation in the vicinity of the pueblo, but Frank's Pueblo itself was classified as an Early Pueblo site, or what is locally termed a Glaze A masonry pueblo, based on the abundance of Glaze A ceramic sherds on the ground surface, and the lack of sherds representing the later glaze ware types. In fact, we found evidence of an earlier adobe pueblo buried under the one with later masonry.

This finding was particularly important because this second adobe pueblo confirms that the concept of the plaza-oriented pueblo was more widespread across Salinas than we previously thought. It also establishes that this kind of site is not associated with any given geographic setting in the uplands or lowlands, and that it predates any social changes that may have occurred with the import of Glaze A ceramics from the Galisteo Basin. Although only these two adobe pueblos have been identified, it is entirely possible that others lie underneath the surface architecture of many or most of the other Salinas masonry pueblos.

At Frank's Pueblo, the underlying adobe pueblo was exposed in excavation units on all four sides of the area that we interpreted as a central plaza. Figure 18 (see chapter 6) shows the location and size of the adobe pueblo. Extrapolating from those walls that we identified, the pueblo is similar to Kite Pueblo in size, shape, and construction. A pit feature (a likely cistern) in the plaza may also date to this early time period. Similar to the situation at Kite Pueblo, later rebuilding in masonry took place in all areas of the pueblo.

The southern room block shows the clearest evidence of this rebuilding. Here, we uncovered two adobe walls perpendicular to one another that defined two rooms of unknown size. The preserved adobe walls represent about 1.4 meters of standing wall above the floor. The adobe walls were later partially rebuilt with masonry to create walls with adobe at the base and masonry at the top. Masonry blocks also filled in an older doorway in one adobe wall, showing that the arrangement of activities in the different rooms and their communication routes also changed over time.

On the southeastern portion of the site, we also uncovered evidence of remodeling in which the adobe walls were extended up and modified with masonry, similar to that seen on the east side of Kite Pueblo. In the same area of the site, more all-masonry rooms were added to existing adobe rooms to create additional roofed space on the perimeter of the pueblo. At Frank's Pueblo, however, these room additions seem to record a rapid influx of population, possibly from surrounding areas. These issues are discussed in more detail in chapters 5 and 6, in the context of regional social relationships. Most relevant here is the discovery that the plaza-oriented adobe pueblo represents a relatively widespread regional expression of some common cultural assumptions and social strategies for organizing people and their daily lives, as played out in the medium of the built environment.

Community Organization in Early Salinas Pueblo Sites

Based primarily on the evidence from Kite Pueblo, we can make some general observations about the organization and operation of the social groups that constructed and inhabited these early adobe pueblos. The overall organization of the built environment provides evidence of community subgroupings, their interactions, and also some specific strategies that people used in constructing the village community as a whole

Evidence of Residential Groups

This evidence from the built environment provides some understanding of the dynamics of community and the size and organization of relevant social groups within the village site. It is clear that the built environment here involves construction of structures at two different social and spatial scales: the individual room and the room block. We do not know what variation there was in room size and room activities for the time period of these earliest adobe pueblos, but the few observed rooms appear to be rather larger than 4 square meters in size. They may measure some 2.5 to 4 meters along one side, about the same interior area noted for the early (large) pit structures. In contrast, jacal structure rooms are typically fairly small in size, typically around 2 to 2.5 meters along a side (Chamberlin 2008).

This difference in room size between Jacal period sites and the adobe pueblos suggests that the concept of "room" and its association with social groups and activities (that is, room function) may have changed over time. Given that the two-room jacal structure is common as the "minimal roofed unit" on jacal sites, it is possible that activities such as sleeping and storage may have been spatially reorganized over time, with these different activities assigned variously to one room or two. At present, however, we do not have the detailed information that we would need to consider room functions for either jacal sites or for the adobe pueblos.

We can, however, argue for a significant change in the social importance of the room block within a village site with the change from jacal structures to plaza-oriented adobe pueblos. Jacal settlements contain many rooms of similar size, but also a variable number of room blocks. The room blocks vary in size, from a minimum of two rooms to an observed maximum of about twenty rooms within each contiguous grouping. However, these very large room blocks represent outliers; more typically, a room block consists of two to six rooms arranged in a variety of ways (Chamberlin 2008). At the early plaza-oriented pueblos, this variation in the size of room block comes to an abrupt end. Instead, the pueblo village site presents a highly standardized arrangement of room blocks that are also highly uniform and standardized in size. In the plaza-oriented pueblo, the standardized room block itself becomes the basic unit of iteration for the site (Rautman 2013a).

Evidence of Social Differentiation

This rather dramatic change in the unit of iteration seen at Kite Pueblo is so strongly patterned that it likely is of considerable social significance. The arrangement of room blocks on pueblo sites is also quite standardized as a line of two rooms across without the presence of "extra" rooms projecting to the east that we saw in the Jacal period. Instead, variation is expressed in room block length. To the extent that these units of the built environment express residential arrangements, it seems clear that the architecture of the room blocks expresses and emphasizes social principles that emphasize a particular idea of group relationships rather than any particular group itself. Individual nuclear families, households, or other social units that may vary in size and composition within a resident group, or across time, would have had to manage the activities of daily life the best they could within the space available.

At Kite Pueblo, we do not have the data that would allow us to test the degree to which these potential social groups were important in structuring people's activities. Neither can we detect the extent to which this uniformity of architecture may have served to de-emphasize or mask existing social differences within and between social groups. What we do see, however, is a new emphasis during this time on the similarity of room block groupings and their close coordination into a highly nucleated village site. In other circumstances or in other contexts, social differences—residential group size, composition, affiliation, or influence—might result in variation in the size and arrangement of each room block. Whether or not these social differences were a significant factor in daily life for the inhabitants of Kite Pueblo, it is very clear that any architectural expression of difference, such as we saw during the Jacal period, ceases abruptly and completely.

I do not mean to imply that social distinctions and differences did not exist among the various households, family groups, or larger social units within the pueblo. In fact, based on our understanding of Jacal period site architecture, it seems clear that as a minimum we might expect to see considerable variation in the size of the social group that would occupy a given room block. What the architecture of Kite Pueblo shows, however, is that these possible social differences were no longer expressed in architectural form. At Kite Pueblo it is obvious that differences in room block size follows organizational principles that emphasize the village site as a whole, rather than expressing or responding to more localized subvillage variation in, for example, lineage size, economic success, or political influence. In contrast to the situation at some other early village sites (e.g., Schachner 2010), room block length in Salinas was apparently very strongly structured by the needs of the community as a whole.

Evidence of Social Integration

Our excavations at Kite Pueblo allow us to make two additional observations about site construction that have significant implications for understanding the social relationships at work. First, the site was preplanned and designed originally as a plaza-oriented pueblo. This site does not represent simply an agglomeration of linear room blocks added to one another, nor is there evidence of temporally discrete series of occupations in different room blocks over time. It is also clear that the enclosed plaza could not have been a later addition to extant room groupings, as is the case in other parts of the Southwest (e.g., Adams 1996:53).

This observation argues that the sites' inhabitants were already well acquainted with one another before the pueblo was built and that they had a clear idea of the size and organizational layout of the village they intended to build. This construction does not express an idea that grew over time; rather, it replaces an existing conception of village spatial organization, clearly and unambiguously promoting village unity while simultaneously downplaying any existing internal variations in the size, composition, and importance of subgroups.

The second important observation was that the overall layout of the site remained relatively constant from the time of its construction and was retained until the time of its abandonment. This observation implies that the idea of relatively standardized room blocks, their arrangement and size that allow construction of a central plaza, and the "closed" layout of rooms surrounding this open space were socially important features—so much so that people retained this spatial organization as much as they could over time.

Evidence of Social Commitment to Place

Chamberlin's (2008) survey results show a trend toward greater settlement nucleation over time, a trend that is expressed primarily through spatial clustering of buildings (room blocks) within a site. This trend is inferred from two observations: (1) the largest room blocks are found only on later jacal sites, and (2) the different room blocks cluster more tightly together on the later jacal sites. However, across the Jacal period, the small two-room room block remains in use (Chamberlin 2008).

The change to plaza-oriented adobe pueblos in the late A.D. 1100s or early 1200s represents a significant reorganization of an extant population into a very different type of nucleated village community. Much of this population may already have been residing in the Salinas area, occupying portions of pithouse sites and possibly portions of jacal sites as well, although it is not at all clear whether all room blocks at the large jacal site were contemporaneous in date. A similar process of aggregation of an extant population into sites characterized by pueblo architecture occurs in the neighboring Rio Abajo region to the west (Marshall and Walt 1984) and in the Sierra Blanca region to the south of the Salinas area (Kelley 1984, 1991; Spielmann 1996). In Salinas, however, there is no evidence of any actual population movement or migration at this time.

Early Adobe Pueblo Communities in Salinas

This detailed study of the construction and occupation of Kite Pueblo provides a slightly different perspective on region-wide social relationships and the significance of the built environment in expressing and developing principles of community organization. The discovery of a buried adobe pueblo of similar size and layout underneath the masonry architecture at Frank's Pueblo confirms that these social strategies were not unique to the inhabitants of Kite Pueblo, but more likely represent ideas that were or became widely shared across the Salinas region. In contrast to the situation in the Rio Abajo, however, there is no evidence that these changing ideas had anything to do with changes in personnel or in regional patterns of social interactions or immigration.

Most importantly, the information obtained from these excavations reveal the inhabitants' concerted and focused emphasis on village community identification and unity, if not control (e.g., Schachner 2001; Graves and Van Keuren 2011). The inward-focused, plaza-oriented pueblo provides an image or metaphor for one specific kind of whole-group dynamic (see Ortman 1998; Potter 2002; Chamberlin 2012), an image and identity that structured not just ritual events but also the daily practice and activities of every member of the pueblo.

In the plaza-oriented pueblo, the built environment creates outdoor space that blurs the distinction between interior and exterior space, between public and private acts, or between communal or solitary events. The central plaza creates outdoor space that is exterior in the sense that there is no roof, yet it is obviously interior at the scale of the pueblo as a whole. Similarly, the space encompassed within the central plaza cannot be identified directly as either public versus domestic, nor as public or communal versus private in use. For example, the central plaza can be a forum for display of shared symbols (Ruscavage-Barz and Bagwell 2006; Chamberlin 2012) as well as a place to conceal one's privately held individual actions and beliefs amid their public expression. Graves and Van Keuren (2011) consider explicitly this concept of the plaza-oriented pueblo as an expression of the ever-present social monitoring of the panoptic gaze (see Foucault 1979). Other outdoor spaces such as rooftops and plaza entryways hold an ambiguous and flexible position as outdoor spaces that are exterior to individual rooms. They, too, are public in the sense that they are potentially visible, yet are not necessarily unrestricted in use.

An additional consideration is that interior space and household tasks cannot be equated directly with domestic or private contexts in village society (e.g., Byrd 1994; Bowser and Patton 2004). Domestic tasks such as obtaining water or firewood may involve use of public space on the plaza; other tasks are often public in the sense of being visible or otherwise obvious to others (the noise of corn-grinding, for example), whether they are performed indoors or out. Similarly, public tasks, that is, those concerning the community as a whole or a portion of the larger group, are often not particularly public at all. Debate concerning community issues might involve many or only a few individuals; these meetings might take place in an open arena or in the seclusion of an interior room such as a kiva.

Whatever the reason that people adopted the plaza-oriented layout, it is clear that in these village sites the built environment would have had a pronounced impact on everyday social interaction and activities. Even if not all rooms were occupied at any given time, the presence of contiguous rooms affects rooftop space, traffic patterns within the site, access between the plaza and residential structures, and visual contact among the site's inhabitants. Furthermore, a limited number of passageways constrain access to the plaza and would also make a person's entry or exit visible to people already on the plaza, or to rooftop observers. In all these ways, exterior space structured social interaction, creating a public social context in which everyday activities, no matter how private or domestic, took place.

The role of visible ritual space such as plazas is an important feature of many societies, including among the ancient and modern Pueblo Indians (e.g., Moore 1996; Schachner 2001; Ruscavage-Barg and Bagwell 2006; Chamberlin 2006, 2012). The information from Kite pueblo argues that the plaza was of particular and continuing importance, an importance not limited to ritual functions or ritually defined time periods. Rather, on a more fundamental level, the centrally located and mutually visible space also plays a role in creation of a shared culture at the scale of the whole village community. In terms of each individual's performance of everyday activities in a public and visible context, the central plaza may have been just as important as the public performance of communal rituals in contributing to the sense of group self-awareness and self-identification that Kolb and Snead (1997) argue is an important aspect of social community.

The plaza-oriented pueblo thus expressed and reinforced in spatial terms an existing, although possibly more fragile, social or political relationship (cf. McGuire and Saitta 1996). In this respect, the plaza-oriented pueblo emphasizes community self-identification and unity as a single group. Much archaeological work among early pueblo groups of the northern Southwest presents a scenario of continuing competition between different groups of people, possibly lineages, within a village site, with different groups experiencing a relatively high degree of subgroup autonomy in organizing and constructing space within the village site (e.g., Schachner 2010). The evidence of early pueblo village organization in Salinas seems quite different. Here, the inhabitants of Kite Pueblo apparently focused on reducing subgroup differences and highlighting village community identity. The plaza-oriented pueblo may have been one strategy that facilitated such self-awareness, not just through the enactment of communal ritual but also through the individual performance of the mundane domestic tasks of everyday life in the public context of the central plaza.

CHAPTER FIVE

Masonry Pueblos and Communities

"Dirt is cool."
—DOMINIC PERRONE (FIELD CREW, 1992)

In the Salinas region, the locations of early masonry pueblos have been known for some time because these sites are relatively large and visible as rubble mounds with some standing masonry. Although there are several early masonry pueblos in Salinas, most of what we know about this time period in Salinas comes from excavations at the very large late pueblos—those that were occupied up to and after the time of Spanish conquest. The list of early pueblo sites in table 13 and those shown in figure 15 include the single component sites and also those where the surface archaeology is dominated by later construction.

At the largest pueblos, however, the early period occupation is buried under standing architecture that represents centuries of subsequent remodeling. Our ideas about the earliest occupation of these sites comes from room fill in rooms exposed during deep and extensive archaeological excavation, such as the National Park Service excavations in Mound 7 at Gran Quivira (Vivian 1979; Hayes et al. 1981). At Gran Quivira Pueblo, Hayes et al. (1981:12) terms this the Early Phase; the presence of Glaze A pottery in these deposits date them from about A.D. 1300 to 1400. Another source of information comes from opportunistic finds during National Park Service salvage excavations associated with maintenance or stabilization projects. Additional information about the early time periods at these large sites also comes from academic research projects in midden deposits that could be identified as Glaze A in date (Spielmann 1992; Graves and Spielmann 1999, 2000a).

Table 13. Pueblo Sites in the Salinas Province

Site Number	*Name*	*Comments*
Pueblo III, adobe pueblos with masonry remodeling		
LA 199	Kite Pueblo	Jumanos Division on the Medanos Plains; southern Chupadera Mesa; small amount of Glaze A pottery associated with remodeling (Rautman 2000)
LA 9032	Frank's Pueblo	Jumanos Division; adobe pueblo component likely pre-dates import of Glaze A ceramics (Rautman, this volume)
The so-called "black-on-white pueblos" (Caperton 1981:6–7) (that is, masonry pueblos that pre-date Glaze A pottery)		
LA 503		Tompiro Division; northern Chupadera Mesa
LA 9029	Pueblo Seco	Jumanos Division; southern Chupadera Mesa; overlooks Arroyo Seco (Chamberlin 2008)
Early Pueblo III "transitional and early masonry" pueblos (Caperton 1981) ("the Glaze A pueblos")		
LA 197p	Montezuma Pueblo	Jumanos Division; southern Chupadera Mesa; includes jacal and also pueblo component (Chamberlin 2008); located near a spring
LA 2091	Pueblo de la Mesa	Jumanos Division near Jumanas Mesa; overlooking Pueblo Colorado; a few Glaze B sherds found
LA 2548		Tompiro Division; near Ténabo
LA 9007		Tompiro Division; northeast Chupadera Mesa
LA 9009		Tompiro Division; northeast Chupadera Mesa; south of the modern town of Mountainair
LA 9012	Lost Pueblo	Jumanos Division; southern Chupadera Mesa; located on a ridge
LA 9014p		Jumanos Division on Turkey Ridge; jacal site and also transitional/early pueblo component (Chamberlin 2008)
LA 9016p		Jumanos Division on Turkey Ridge; jacal site and also transitional/early pueblo component (Chamberlin 2008)
LA 9021	Axtell Pueblo	Jumanos Division; southern Chupadera Mesa; non-defensive location
LA 9026p		Jumanos Division; southern Chupadera Mesa (Chamberlin 2008)
LA 9032p	Frank's Pueblo	Jumanos Division; southern Chupadera Mesa; pueblo component

Table 13. (continued)

Site Number	*Name*	*Comments*
Pueblo IV late masonry pueblos		
LA 51	Pueblo Blanco (Tabirá)	Jumanos Division near Jumanas Mesa; Glaze A through Spanish occupation; Catholic visita (chapel) associated with Abó
LA 83	Pueblo Pardo	Jumanos Division near Gran Quivira; Glaze A until Spanish occupation; Toulouse and Stevenson (1960:3) date abandonment to 1630
LA 95	Quaraí	East Tiwa Division; Glaze A time through Spanish occupation; Catholic mission site until 1677
LA 97	Abó	Tompiro Division; Glaze A through Spanish occupation; Catholic mission site until about 1673
LA 120	Gran Quivira	Jumanos Division; Glaze A through Spanish occupation; Catholic mission site until 1671
LA 200	Ténabo	Tompiro Division near Abó; Glaze A until Spanish occupation
LA 476	Pueblo Colorado	Jumanos Division near Jumanas Mesa; Glaze A through E ceramics present; abandoned before Spanish occupation
LA 2541		Tompiro Division; near Ténabo; Glaze A to C
LA 9008	Gypsum Spring Pueblo	Tompiro Division; northeast Chupadera Mesa; small late masonry pueblo; Glaze E pottery only
LA 9043		Tompiro Division; northeast Chupadera Mesa; small late masonry pueblo; Glaze D and E pottery only

Note: The dates of abandonment of Abó, Quaraí, and Gran Quivira are from Ivy (1988:233). Unless otherwise stated, all other comments are based on Caperton (1981:6–8). Piro Division (Mera 1940) pueblo sites on Chupadera Mesa are not included here (see Mera 1940; Kyte 1988; Montgomery and Bowman 1989).

In this chapter, I consider evidence of early occupation at two early masonry pueblos where occupation did not continue into later time periods: Pueblo de la Mesa (LA 2091) and the masonry occupation of Frank's Pueblo (LA9032). These sites are small enough that it is possible to get a sense of the pueblo site as a whole, the constructional sequence, and people's strategies of remodeling and modifications over time. This detailed constructional history provides more of an understanding of how people actually lived within this built environment, the ways in which people interacted on a day-to-day basis, and the social processes involved

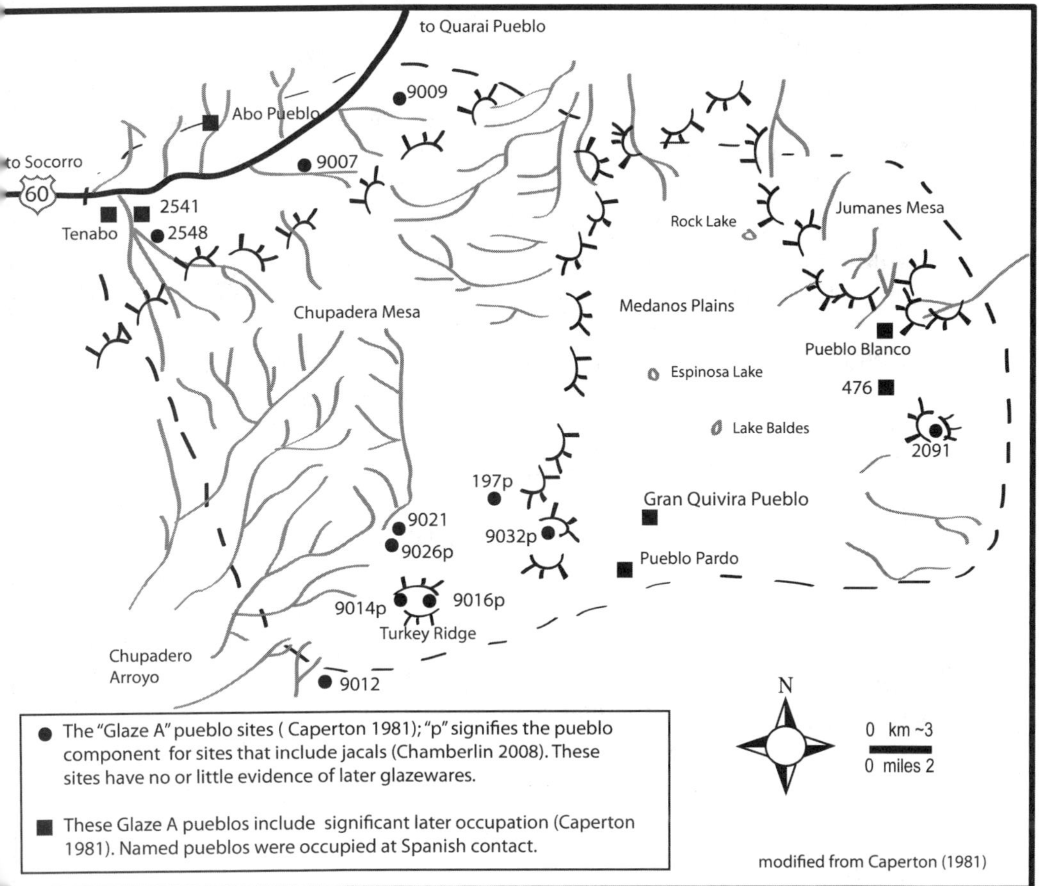

Figure 15. Location of the Glaze A pueblos in the Salinas Province. Modified from the base map shown in Caperton 1981:xii.

in village life. These sites also represent the last of the archaeological village sites that we can reasonably describe as early village communities in the sense defined here.

These two sites are located in two different subareas of the Salinas region. Pueblo de la Mesa is located in the eastern portion of the Salinas region amid a spatial grouping with two other large sites that H. P. Mera described by Jumanes Mesa (Mera 1933, 1940; see figure 15). Jumanos Mesa represents large spur or eastward projection from the main part of Chupadera Mesa; in some reports this area is termed the "East Mesa."

Nearby large sites include Pueblo Colorado and Pueblo Blanco (Tabirá). Both of these sites were occupied during the Glaze A period and also for many years thereafter; they grew to include hundreds of rooms and multiple plazas. Only Pueblo Blanco shows signs of Spanish occupation; Pueblo Colorado was presumably abandoned before Spanish contact (Stubbs 1959).

Pueblo de la Mesa is particularly important for understanding local cultural history partly because a deep extramural midden yielded a great deal of information about daily life during this time period (Rautman 1992, 1993a). In addition, a small population of human burials was found within room fill (see Colyer 1996). Comparison of the bone chemistry information from Pueblo de la Mesa's population with that from the skeletal population at Kite Pithouse Village made it possible to establish that basic dietary composition remained fairly constant across time, from the Pithouse period through Glaze A (Katzenberg and Kelley 1991; Colyer 1996; see related discussions in Whalen 1981a, 1981b, Rocek 1995, and Hegmon 1996 for other regions).

The second early masonry pueblo, Frank's Pueblo, had an adobe component, which was discussed in chapter 4. However, most of our excavated data come from the later masonry occupation. In this chapter and the next, I describe how excavations at these two sites contribute in very different ways to our understanding of early village society and community organization in the Salinas Province, beginning about A.D. 1300.

Early Masonry Pueblos in Salinas

At about A.D. 1300, people in Salinas invested in significant remodeling and new construction on or near their previous village sites. For some reason, adobe or jacal was no longer the building material of choice; rather, people turned to stone masonry for construction. We do not know if the remodeling that we see on these archaeological sites indicates rebuilding after an intervening period of non-use, that is, site abandonment and reoccupation, or if the residents simply reassessed the values of masonry construction for maintenance and remodeling. In either case, the two known adobe pueblos—Kite Pueblo and the adobe pueblo at Frank's Pueblo—experienced a similar sequence of remodeling over time, with stone masonry replacing adobe block construction. What we do not know, however, is how many other of the known masonry pueblos may have experienced a similar constructional history.

When people did change to masonry construction, they apparently did so using a variety of construction strategies. At both sites, for example, some adobe rooms or walls were partially or completely razed and rebuilt with masonry. At each of these sites at least some of the new masonry rooms were constructed as additions outside the pueblo, abutted to the old adobe walls. At the same time, people made some other changes in the built environment. For example, they altered the size of some rooms (shown at Kite Pueblo) and also the patterns of access between rooms (shown at Frank's Pueblo).

At both sites, this adobe to masonry remodeling effort occurred at about the same time that people in Salinas began to import Glaze A pottery from the northern Rio Grande. Toulouse and Stephenson (1960) termed this time period the Gran Quivira Focus, but this nomenclature has not been widely adopted except as reported in summaries by Stuart and Gauthier (1981). Instead, people refer to "early masonry pueblos" (e.g., Caperton 1981) or "the Glaze A period pueblos" (Kyte 1988, 1989; Montgomery and Bowman 1989).

This association between masonry construction and the presence of Glaze A pottery is so close that only two known pueblos deviate from this pattern. In the archaeological literature they are sometimes called the "black-on-white pueblos" or "black and white communities" (e.g., Mera 1940). These terms refer to the dominance of Chupadero Black-on-White pottery and also the lack of Glaze A and later glaze sherds from surface collections. Because glazed pottery sherds have not been found on these sites, archaeologists have assumed that these two sites were built before about A.D. 1300 (Mera 1940; see also Kyte 1988, 1989; Montgomery and Bowman 1989).

LA 9029 (Pueblo Seco) in the southern portion of Salinas is one of these so-called black-on-white pueblos, as is LA 503 near Abó Pass. These two pueblos are also considered unusual for other reasons, including their location, apparently rapid construction, and short duration of use. They are both located in upland locations, on high promontories where access is difficult. Unlike the other known masonry pueblos in Salinas, there is no evidence of any prior Jacal period occupation in these locales; they seem to represent new occupations. They were both apparently constructed fairly hastily and were occupied for only a short period of time. These pueblos have therefore been interpreted as special purpose "fortress sites" (e.g., Caperton 1981). I will discuss these so-called black-on-white pueblos in the next chapter, in the context of interpueblo relationships within the Salinas region.

In general terms, however, the Glaze A pueblos include the small early masonry pueblos as well as the early pueblo occupation at the larger late sites such as Abó, Quaraí, and Gran Quivira (see table 13). These Glaze A sites may not be particularly large, but local ranchers have known about them for many years because the collapsed masonry walls create clearly visible and also durable mounds of rock and sand.

Glaze A Pottery and Early Masonry Pueblos

The presence of glazed pottery on these small pueblo sites is important not only because development and use of glaze paint represents a significant change in ceramic technology but also because its import into the Salinas area coincides with some other widespread social changes. These social changes include a period of regional abandonment in the northern Southwest around the Four Corners and the Colorado Plateau, and large-scale regional population movements southward, along the Rio Grande. Associated with all these population movements are changes in people's social institutions, the organization of religious and cultural traditions, and people's view of themselves and their place in the world (Habicht-Mauche 2006).

The manufacture of glazed pottery involves new techniques of combining raw materials into the glaze paint and also involves a number of other changes in the way that the pottery is fired and the vessels are formed and decorated (Shepard 1942, 1954; Warren 1981). H.P. Mera (1935, 1940) distinguished temporal changes in these techniques and proposed a chronological sequence (Glaze A through Glaze F) based on these changes. Mera's system of classification considers the character and composition of the glaze and also the shape of the vessel rim (see also Shepard 1936 and Hayes et al. 1981 for specific information about Salinas glaze wares). Despite much later research, Mera's classification system is still the most commonly used today (Habicht-Mauche et al. 2006). Table 14 lists the generally accepted dates of production for each glaze type, but there is debate about the extent to which the different ware types overlap in time and the extent to which a given type should be used as a chronological marker (Snow 1973, 2007; Eckert 2006a, 2006b).

Adoption and use of glazed pottery also signals regional changes in the way that ceramics were exchanged and used among the Rio Grande pueblos and across the Pueblo Southwest in general (Habichte-Mauche 1993, 2006; Creamer 1996, 2000). For example, we know that glazed pottery was

Table 14. Production dates for Rio Grande glaze wares

Glaze ware	*Production dates**	*Named types*
Glaze A	A.D. 1313 to 1500+	Los Padillas Glaze-on-Polychrome; Agua Fria Glaze-on-Red; San Clemente Glaze-on-Polychrome; Pottery Mound Glaze-on-Polychrome; Cienequilla Glaze-on-Polychrome
Glaze B	A.D. 1410 to 1500+	Largo Glaze-on-Yellow; Medio Glaze-on-Polychrome
Glaze C	A.D. 1430 to 1600+	Espinosa Glaze-on-Polychrome
Glaze D	A.D. 1460 to 1550+	San Lazaro Glaze-on-Polychrome
Glaze E	A.D. 1480 to 1630+	Puaray Glaze-on-Polychrome; Pecos Glaze-on-Polychrome
Glaze F	A.D. 1525 to 1700	Kotyiti Glaze-on-Red and Glaze-on-Polychrome; Jornada Glaze-on-Red and Glaze-on-Polychrome

*Production dates are from Eckert (2006b:37); see Hayes et al. 1981:table 120, p. 92 for further information regarding the chronology of the different named types.

particularly important in Pueblo Indian ritual in public feasts and in inter-pueblo social interactions and political negotiations, where they assumed an important role as symbols of economic and ritual power and social influence. In particular, the large glaze ware bowls (those about the size of a modern formal punch bowl or salad serving bowl) were important in public feasts for serving and display (Spielmann 1998b, 2002; Mills 2007).

We know that people also used similar large bowls in specific social contexts into the nineteenth and twentieth century. For example, the ethnographer Matilda Coxe Stevenson (1904) described Zuni women carrying similarly large bowls filled with food to community events, and filling bowls with leftover food to send home with visitors. Typically, these large bowls were decorated on both the inside and the outside; the implication is that even when filled with ground maize meal or other foodstuffs, the boldly decorated outsides of each bowl would be highly visible to participants and onlookers in these social events (see discussion of viewer distance and symbolic communication in Chamberlin 2002, 2006, and 2008).

Large jars with glazed decorations are also important indicators of this time of widespread social changes after the development of Glaze A pottery and may indicate the spread of ritual beliefs and the ritual use of specific items. Chamberlin (2002) examined the physical characteristics of these vessels and concluded that they were likely manufactured for ceremonial storage, rather than for storage of foodstuffs. He found that characteristics of the clay such as porosity are similar to that described ethnographically for vessels used to store ceremonial water. The use of glaze ware jars is also associated with dry storage for seeds and also non-food items used in rituals (Chamberlin 2002).

In the research discussed here, I use glaze ware pottery primarily to identify the date of site occupation. One problem is that people continue to make Glaze A pottery over a long period of time, even as they are also making and using Glaze B or Glaze C vessels. When archaeologists identify a pueblo as Glaze A, therefore, they are saying that they found only Glaze A pottery at that site.

Excavations at Pueblo de la Mesa (LA 2091)

Pueblo de la Mesa is recorded as a Glaze A/B pueblo, indicating that some Glaze B sherds were found on the ground surface when the site was first identified (Mera 1935). Throughout fieldwork at the site, however, I found only Glaze A sherds in excavated contexts and only a small number of sherds of Glaze B pottery on the ground surface. Sherds from a Glaze C vessel were found only in association with a late burial in Room 5, described below. This information suggests that regular occupation of this site ceased just after A.D. 1425 or thereabouts, when Glaze B pottery started to be produced in the Rio Grande. This expectation was confirmed by radiocarbon dates from the site (table 15).

The site is located on the northernmost spur of a small but locally obvious isolated mesa in the area around Jumanas Mesa (see figure 15). There is a marked drop-off on the northern portion of the site, but the mesa slopes more gently down to the south. The sloping portion of the mesa consists of Glorieta sandstone; the more vertical portion near the top marks the contact with the more weather-resistant overlying San Andres limestone. This mesa-top situation affords a good view of the landscape to the north and east and overlooks Pueblo Colorado (LA 476) about 4.8 kilometers to the northwest. The second largest late pueblo in this area, Pueblo Blanco, or Tabirá (LA 51), is also located on the sandy lowlands, about 10 kilometers distant.

Mera (1935) originally suggested that Pueblo de la Mesa was a fortified refuge associated with the early occupation at Pueblo Colorado, an

Table 15. Radiocarbon dates for Pueblo de la Mesa, LA 2091

Context	*Radiocarbon (B.P. ± 1σ*	*C13/C12 ratio**	*C13 adjusted age*	*Calibrated date (± 2 σ)*
Kiva 1	150 ± 50 B.P.	–2.8 ppm	515 ± 55 B.P.	A.D. 1310–1350; A.D. 1380–1460
Room 5	375 ± 50 B.P.	–8.3 ppm	650 ± 55 B.P.	A.D. 1270–1410
Room 3	240 ± 50 B.P.	–10.2 ppm	480 ± 55 B.P.	A.D. 1400–1500
Room 4	560 ± 60 B.P.	–10.6 ppm	790 ± 60 B.P.	A.D. 1160–1300
Cistern	520 ± 60 B.P.†	–12.6 ppm	720 ± 60 B.P.	A.D. 1220–1400

*The ratio was measured in the accelerator and may include an instrument component. Typical ratios for maize are -8 to -12 ppm.

†The sample contained less than 1 g of carbon and was given extended counting to increase precision.

interpretation later reported by Stuart and Gauthier (1981:table VII.11). My excavations at Pueblo de la Mesa in Cibola National Forest took place in the early 1990s, partially as a salvage excavation in response to an observed increase in illegal pot-hunting activities. One of the research objectives for this study was to test whether it was in fact an ordinary living site or whether it represented a special-purpose fortress or refuge.

There are several lines of evidence that might help distinguish site function (e.g., Adams 2004, or more generally, Newman 1972). For example, one might expect a fortress/refuge to have been built in its entirety fairly quickly and to include defensive walls limiting access to the site. It might have a number of interior or defensible storage facilities for food and water (or disproportionate space allocated to storage), caches of weaponry, and (in general) to exhibit less evidence of the normal activities of daily life, such as pottery manufacture, stone tool manufacture, or basic food processing (Adams 2004:1–7). According to these criteria, I concluded there was little evidence that Pueblo de la Mesa was specifically or primarily a fortress or refuge. However, my later excavations at Frank's Pueblo in 2007 and 2008 suggest that defensive strategies at Pueblo de la Mesa may have been more important than I initially appreciated. While I still argue that Pueblo de la Mesa was not exclusively or primarily a fortress retreat, it is likely that the threat of interpueblo conflict in the region was indeed a factor in site selection and use during this time. In the rest of this chapter, I discuss how excavations at this site contribute to our understanding of early masonry pueblos as living village communities within this regional context.

Site Organization and Construction

The overall shape of Pueblo de la Mesa is similar to a figure "8," with the long axis of the "8" oriented to the northeast (figure 16). The backbone of the site is a relatively high (1.5 meters or so) rubble mound oriented northwest to southeast, representing rooms that may have been two stories in height. That is, we excavated about 1.5 meters down to the floor in these rooms, and there was still considerable rubble on the ground surface. The amount of observed rubble, added to the standing wall height, would be more than sufficient to represent a second story.

All other room blocks at the site are likely one-story construction. These structures surround two plazas, which are represented by the empty spaces within the figure "8" loops. In addition, a small but undefined number of jacal structures border the exterior perimeter of the pueblo on the south and west, and are obviously connected to the pueblo itself. I suggest that these jacal structures might be more accurately envisioned as porches or sunshades (ramadas) rather than enclosed living rooms or evidence of early Jacal period occupation.

These excavations showed that the pueblo was constructed in a series of different phases, a pattern very different from that observed at Kite Pueblo. The visible architecture of the site thus records a process of accretion and growth over time, rather than a single construction episode.

Construction of the visible architecture at this pueblo began with a long room block (Roomblock II). This room block was two rooms wide and about eight or more rooms in length. It was constructed in one episode, using relatively thin slabs of sandstone. The long walls were built first, and the interior partition walls were then added using ladder construction.

This original long room block and a smaller one to the northeast defined the North Plaza. The smaller room block (Roomblock I) had a complicated history of remodeling. The first rooms built in this area of the site were quite large, but of unknown size. The north and west walls of one excavated room (Room 5) were added later, partitioning the early large room into a smaller one. At least three floor surfaces were defined in this room, representing episodes of remodeling. On the uppermost floor was a stone-lined hearth. We also found a burial, an adult female, whose body had been placed on the floor next to the hearth. No obvious burial pit could be identified, but sherds of a late glaze (Glaze C) vessel were found near the body, suggesting that the burial was placed into room fill after the main period of regular site occupation.

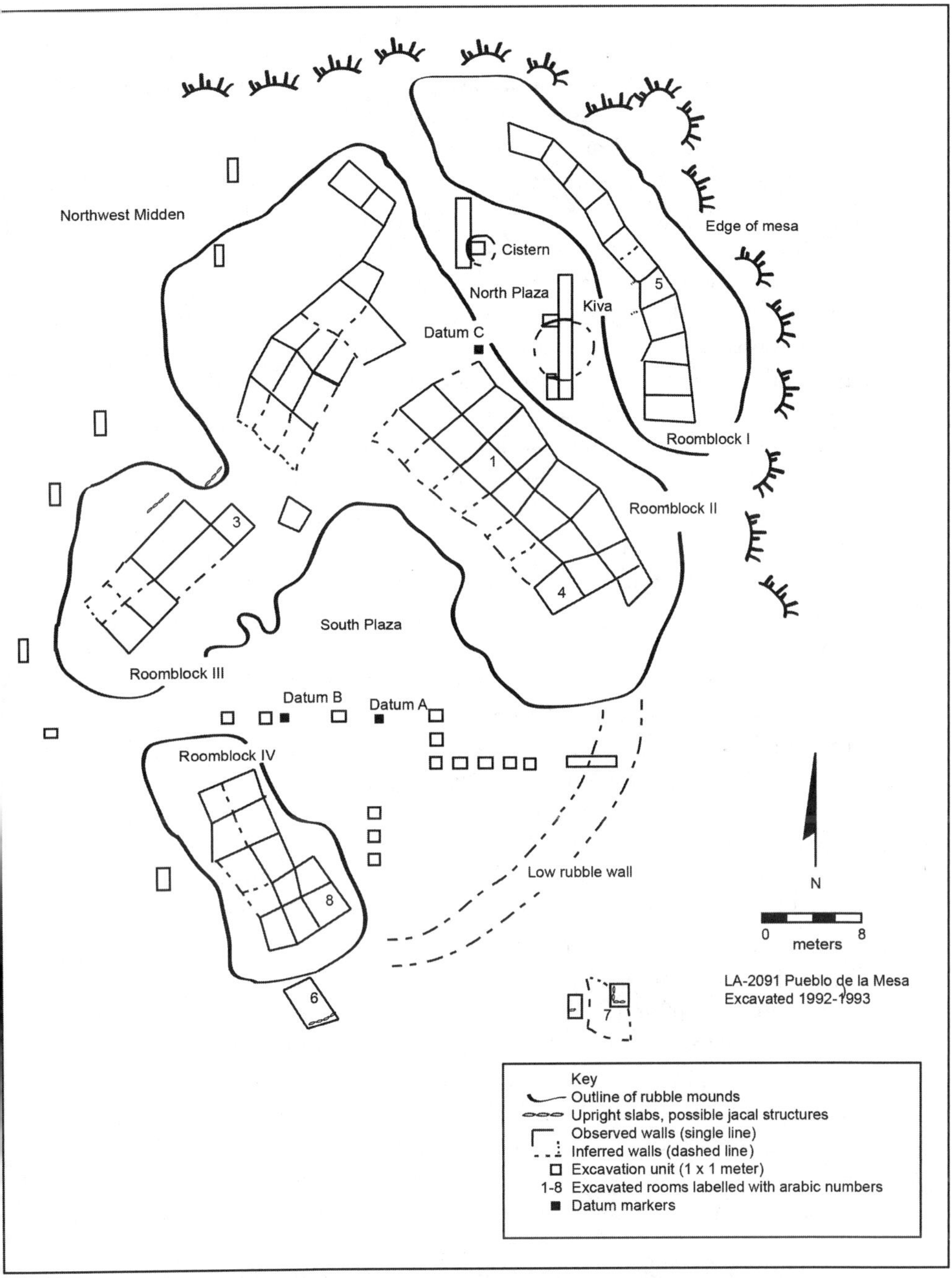

Figure 16. Map of Pueblo de la Mesa, LA 2091.

The North Plaza also contained two circular structures. Both structures had been dug into limestone bedrock, with interior height increased by the addition of limestone masonry walls on top of the bedrock surface. The inhabitants filled the base of both structures with hard-packed clay or adobe to make a level surface at the base.

My earlier reports and notes refer to Kiva 1 and the smaller Kiva 2. Kiva 1 represents a large circular semisubterranean ceremonial structure, and a well-preserved one at that. Two different colors of adobe plaster on the interior walls marked successive episodes of wall maintenance. The remnants of a large upright wooden beam adjacent to the interior perimeter wall suggest that the structure once included a roof. In contrast, further excavation showed that "Kiva 2" was really very small in diameter (less than 2 meters) and showed no interior features. In retrospect, I now consider that this pit was more likely to have functioned as a cistern or water storage feature within the plaza, similar to those recorded at Kite Pueblo and also at Gran Quivira.

Construction of both features, Kiva 1 and the proposed cistern, would have involved removing a considerable volume of limestone. It is therefore feasible that these pit features supplied the distinctive blocky limestone that people used in the second phase of room construction. Certainly there is no local scarcity of available sandstone on the sides of the mesa, if the builders chose to access it. If, however, the site's inhabitants were also interested in digging reservoirs or ceremonial chambers, certainly utilizing the resulting limestone blocks for related aboveground construction represented a sensible strategy.

Other interesting cultural features were located on the north side of the site. In this area, the contact between the San Andres limestone stratum at the top of the mesa and the underlying Glorieta sandstone is located only about two meters below the rim of the mesa, just a couple of meters from the pueblo itself. Several sandstone outcrops are exposed on the north rim of the mesa and some have been culturally modified. In at least two locations, the upper surface of the exposed sandstone has been worn away, as if someone used it as a grinding surface to polish a tool (e.g., the size of a stone axe head). Several of these worn areas are trough shaped, similar to the work surface of a trough metate; at least one area has been utilized with a circular, overall motion to create a shallow, round basin-like depression.

South of this area, three detached room blocks define a larger plaza, the South Plaza. On the southeast side of the site, a knee-high rubble wall made of very large unshaped limestone blocks curved around and enclosed the South Plaza.

Farther south, outside the pueblo, several features appear to be reservoirs or agricultural features. The features are either completely circular or arc-shaped; the circular ones are typically about five meters in diameter. Similar kinds of circular reservoir features are recorded at Gran Quivira (Toulouse 1945) and are still visible today. If these shallow bowl-shaped pits were plastered with adobe or clay, like the clay-lined pit at Kite Pueblo, they could have retained a substantial supply of rainwater or at least slowed its runoff. The features that represent incomplete circles are of roughly comparable size and are also interpreted as water control features or some sort of terraced agricultural feature. The stone portions of the arcs are always oriented at right angles to the local slope and with the concave portion upslope, a configuration that is consistent with use for water retention. Not surprisingly, these arc-shaped features also show accumulation of sediment on the upland side of the arc. A dozen or more of these features extend south and west of the site, to a distance of about one hundred meters from the pueblo.

Because access to water is such a recurring issue for people living in the Salinas region, it is worth noting here that Caperton (1981:9) mentions the existence of a spring about two and a half kilometers south of Pueblo Blanco. This location would place the spring about halfway between Pueblo Colorado and Pueblo Blanco and within walking distance of Pueblo de la Mesa as well (roughly eight kilometers). Caperton (1981:9) reports that the spring went dry in the 1940s and could not be located at the time of his survey in 1979.

Occupational Sequence

Pueblo de la Mesa showed evidence of two overlapping phases of occupation, which we can infer from the sequence of room construction and abandonment (e.g., Cameron 1991, 1999a; Bradley 1993; Cameron and Tomka 1993). The early phase coincides with construction of first Roomblock II and later the visible architecture of Roomblock I; in this time period, the North Plaza was defined. A later period of construction and occupation is recorded in Roomblocks III and IV, which together ring the South Plaza. When these two room blocks were occupied, the rooms around the North Plaza were abandoned and the rooms filled with household debris and windblown sand.

When Roomblocks I and II were in use, the North Plaza was well defined on the south and northeast, funneling foot traffic into the plaza from its western side. The tall back wall of Roomblock II limited access to the pueblo from the south, where the ground was gently sloped. On the north and east sides of the pueblo, the steep sides of the mesa would discourage access to the North Plaza. At this time, low areas of the bedrock on the North Plaza were built up with adobe to make a level surface. Activities such as tool manufacture and food processing produced midden deposits including abundant maize as well as remains of deer, pronghorn, and rabbits.

At this site, a change in the stone used for construction also marks the passage of time. The earliest rooms (Roomblocks I and II) were constructed with walls made of relatively thin slabs of sandstone set on limestone bedrock foundations. Construction during this second period is differentiated by the use of blocky limestone for construction and a shift in occupation focus to rooms surrounding the South Plaza. The two pit features in the North Plaza, dug into the limestone cap rock, may be associated with this second period of building activity.

At this time, a new line of one-story rooms with walls made from blocky limestone was added to the southwest side of Roomblock II. These new rooms bordered the South Plaza. It is not clear whether this line of rooms represents an addition that encroached onto the South Plaza, or whether the rooms predate the definition of this plaza. The limestone used for this addition could have come from construction activities on the North Plaza, which involved removal of a considerable amount of limestone to form the two pit features on the plaza. The exact temporal sequence of these activities, however, is not clear.

Based on the similarity of construction (the use of blocky limestone), however, it seems likely that this addition occurred around the same time as the construction of later room blocks to the west (Roomblock III) and south (Roomblock IV). If so, then this period of construction represents an increase in the overall size of the resident community. At the same time, there was obviously an effort to coordinate construction around a new central plaza. Several jacal surface structures abut Roomblock IV on the outside of the pueblo to the west and also form a small freestanding grouping of structures to the south that we called Room 7. These jacal structures are most likely porches or ramadas associated with the main period of site occupation. There is no evidence that they might date earlier in time.

The sequence of abandonment of the rooms and room blocks at Pueblo de la Mesa follows a broadly similar north-to-south timeline. The kiva, cistern, and room blocks and rooms around the North Plaza and on the northern portions of the South Plaza were filled with midden sediments, including organic and inorganic domestic debris. In addition, the area west of the pueblo itself yielded abundant midden debris, including large quantities of deer and pronghorn bone, as well as the remains of smaller animals. In contrast, the south side of the site was the last portion to be abandoned. The near-zero artifact content of the rooms of Roomblock IV indicates that this room block probably housed the last people who lived at the pueblo. After they left, these rooms filled with windblown sand.

Evidence of Daily Life at Pueblo de la Mesa

Pueblo de la Mesa represents a particularly rich source of information regarding subsistence and material culture due to the well-preserved, thick, and dense extramural midden deposits located just west of the site, as well as within several of the rooms. Excavated plant and animal remains indicate that people in this pueblo used domesticated maize, beans, and squash, and also non-domesticates, including twenty species of wild plants and a dozen genera of animals (table 16). There is evidence of ceramic manufacturing at the site itself, and it is also possible that ground stone tools were manufactured here from the local sandstone. These activities point to a long-term resident population of people participating in farming, hunting, making ceramic vessels and stone tools, grinding maize, processing wild plant foods, and engaging in other activities of daily life. Some of the details of these activities can be reconstructed from the abundant artifact content found in middens and in room fill, and on room floors and in hearths. These details are discussed briefly below.

Animal Remains

Faunal remains record that the inhabitants hunted pronghorn, mule deer, and the occasional bison; they also used a number of smaller animals, particularly cottontail rabbit, pocket gopher, jackrabbit, and prairie dog. Domesticated turkeys were also present; in the ancient Southwest, turkeys were kept for their feathers as well as food, and their bones are sometimes used for tools (McKusick 1986). Abundant animal bone fragments occur in every excavated context, but are most concentrated in the midden immediately west of the pueblo, in room fill and in the possible cistern. The two plazas, although presumably used at different times, are remarkably similar in terms of the density and overall composition of species recovered and hence were probably the loci for similar sorts of activities.

Table 16. Plant and animal taxa identified at Pueblo de la Mesa, LA 2091

Animal Taxon	Common Name
Bison	Buffalo/bison
Antilocapra cf. *americana*	Pronghorn (antelope)
Odocoileus spp.	Deer
Artiodactyl	Deer or Pronghorn
Sylvilagus spp.	Cottontail rabbit
Lepus spp.	Jack rabbit
Neotoma spp.	Wood rat
Cynomys spp.	Prairie dog
Spermophilus spp.	Ground squirrel
Geomys spp.	Pocket gopher
Peromyscus spp.	Mouse
Lynx rufus?	Bobcat?
Meleagris	Turkey
	Turtle shell

Plant Pollen Taxon	Common Name
Arboreal pollen	
Juniperus spp.	Juniper
Picea spp.	Spruce
Pinus spp.	Pine
Quercus spp.	Oak
Salix spp.	Willow
Nonarboreal pollen	
Chenopodium spp. *Amaranth* spp., *Tidestomia* spp.	Includes goosefoot and pigweed
Cleome	Beeweed
Compositae	Sunflower family
Artemesia	Sagebrush family
Liguliflorae	Dandelion and chicory
Brassicaceae	Mustard family
Ephedra	Mormon tea
Eriogonum	Wild buckwheat
Fabaceae	Bean and legume family
Hydrophyllaceae	Waterleaf family
Lamiaceae	Mint family
Leptodactylon	Prickly-phlox
Nolina	Beargrass

Table 16. (continued)

Plant Pollen Taxon	**Common Name**
Onagraceae	Evening primrose family
Cylindopuntia	Cholla cactus
Opuntia	Prickly pear cactus
Poaceae	Grass family
Portulaca	Purslane
Rosaceae	Rose family
Cercocarpus	Mountain mahogany
Rumex	Dock
Sphaeralcea	Globe mallow
Typha	Cattail
Cucurbita	Squash, gourd
Zea	Maize

Data from Cummings 1995

All of the animals identified at the site can be found in the area today, with the exception of bison. Bison may have been obtained locally, but the dominance of rib bones present at the site suggest that it was hunted or obtained by exchange from some distance away and brought to the site in dried (jerked) form. Regardless of how the bison was obtained, it is obvious that most of the processing of the animal took place elsewhere, the bulky and heavy bones discarded, and just the meat itself with a few attached bones brought into the pueblo. On the other hand, grazing animals such as mule deer and pronghorn are represented by a wide variety of body parts scattered in many different contexts. Elk are not unknown in the area today, and a local volunteer excavator did note elk tracks on the site one morning, but no archaeological elk remains were identified at the site.

The most common small animal represented in the various cultural deposits is rabbit, particularly cottontail. Many different skeletal elements of both cottontail and jackrabbit occur at the site, indicating that these animals were probably obtained locally and brought back to the village intact for processing.

Use of Plants

Wood charcoal at the site was predominantly juniper, with piñon pine as a secondary resource. Oak, saltbush, and sagebrush were minor species. Only one piece of more exotic ponderosa pine (*Pinus ponderosa*) was identified. Ponderosa pine is commonly used for pueblo roof beams, and good ones are not necessarily easy to find. For this reason, people often reused the larger roofing timber in pueblo construction. Even with some timber reuse, however, increased construction of the largest pueblos must have taxed the local supply. Ponderosa pine is sensitive to not only climatic conditions but also fire regime and other highly variable factors; this species of tree is also relatively slow to reestablish stand density after logging (White 1985). Today, ponderosa pine is found only in the Manzanos Mountains, and in a small relict grove on top of Jumanes Mesa. It seems likely that the pueblo's inhabitants were reserving the scarce ponderosa wood for construction rather than for fuel.

As might be expected at a pueblo site, domesticated maize was ubiquitous in every context. The distribution of different parts of the corn plant across the site, however, provides more information about how this food plant was used and where different activities took place. Corn pollen is most abundant in the hearths, suggesting that much of the work of corn husking and processing took place indoors. Pollen aggregates in the Northwest Midden and the kiva may represent bulk discard of husks containing corn silk; this proposed pattern of disposal reinforces the idea that husking likely took place at the site itself, rather than in the fields (Trigg 1995). The most numerous plant parts recovered across the site were maize cupules, the part of the cob that holds the kernels. These probably represent the use of cobs for fuel. When cobs are burned, the tough fibrous cupules are more likely to be preserved in the archaeological record than the rest of the cob (Trigg 1995).

Use of Stone for Tools

Chipped stone remains at the site indicate a relatively expedient use of stone, which produces numerous unutilized flakes and few formal tools. The formal tools found are primarily scrapers and bifaces—tools that may have been most likely used in processing of animal resources or coarse vegetal matter.

The ground stone assemblage included tools such as manos and metates, and also thin slabs, about two centimeters thick, made of either sandstone or limestone. Similar slabs are often used in pueblo sites for defining mealing bins or as doors (e.g., Hayes et al. 1981). Grinding tools at Pueblo de la Mesa were made nearly exclusively of sandstone, which was of course readily available just downslope in the immediate site area. Other stone present included vesicular basalt lava, which had to be imported to the area. This kind of lava likely derives from one of the many lava flows that occur along the Rio Grande Rift (see Chapin and Seager 1975).

Evidence of Ceramic Use and Manufacture

Ceramics were abundant in all portions of the site, but particularly in the two plazas and in the midden just west of the pueblo (Rautman 1995c). The ceramic sherds derive mostly from white ware and brown ware jars, with a small number of glazed red ware bowls. This sort of assemblage represents a variety of activities, from water storage (the white ware jars) to cooking (brown ware utility jars). The large glaze ware bowls are usually interpreted as serving bowls that would be used in relatively formal contexts, rather than in ordinary domestic chores.

Local pottery such as Corona Corrugated brown wares and Chupadero Black-on-White are the most common pottery types. A rather spectacular collection of three unfired ceramic vessels on the floor in Room 1 indicates that ceramic manufacturing took place here also. These vessels were found grouped closely together near the northeast corner of the room, crushed flat under the weight of the overlying sediments. Two of the vessels were very poorly preserved, but the third was obviously a round bowl, measuring about twenty centimeters across, that had been set upside down (orifice down). A second bowl was apparently placed on the floor, orifice up. All were made from a reddish brown clay; at least one of the vessels showed remnants of a white slip on its exterior surface. A collection of sherds of unfired clay found underneath the third vessel suggests that a fourth vessel may also have been present. The bowls were apparently formed, slipped, and set to dry before firing, suggesting that they were in fact manufactured on-site.

These clay vessels sat upon an adobe floor surface that had been built directly atop bedrock. A second, upper floor was constructed later, forty centimeters above the first one. The fill between the two floors contained burned clumps of adobe and many burned pieces of sandstone rubble; it was loosely packed, with many air pockets around the rubble.

The evidence suggests that people left their work in progress rather abruptly and possibly unexpectedly. The burned sandstone suggests that fire played a role in the history of this room, but we do not know if people had to abandon their work in progress because of the fire or whether the room experienced burning at some later time after the people left. We do know, however, that after the fire, people returned to the room and without clearing the rubble all the way down to the original floor, constructed a new floor surface and hearth on top of the compacted burned debris.

Interpreting the Significance of Fire Damage in the Pueblo

This evidence indicates that fire played a role in affecting occupation and reoccupation of various portions of the pueblo. In addition to the burned rubble observed in this room described above, other rooms exhibit thick black soot on portions of the standing walls, and/or fire-reddened masonry walls and bits of rubble.

It is difficult to interpret how extensive or damaging these fires really were (Rautman and Chamberlin 2010). Today, fire fighters can often determine the source of a fire, the pattern of the burn, and investigate the use of accelerants to distinguish a natural fire from arson. Archaeologists do much the same sort of detective work in interpreting burned structures. For example, an experimental archaeology study of burning in a wattle-and-daub structure showed that a hearth fire might spread to a thatched grass ceiling. The most extensive damage in this case was to the organic roofing material (Bankoff and Winters 1996). Other experimental studies at the archaeological site of Homol'ovi, near the Hopi Mesas in Arizona, provide further evidence for evaluating fire damage within masonry or adobe structures. Adams (2004) found that in a pueblo context, artifacts on the floor were little affected by the heat from an interior fire, while the uppermost portions of the walls sustained the most damage from flame, smoke, and heat.

The Homol'ovi experiments also show that pueblo room fires tend to be self-limiting. The thick roof beams do not ignite easily from a normal interior source of flame such as a cooking fire. In addition, a pueblo's adobe roofing would also limit the spread of the fire from one room to another. In the event that a fire within one room did burn through the beams, the roofing material (including adobe) would fall into the room, smothering the flames, while the masonry walls generally would keep the blaze confined within the room (Adams 2004). In other words, while one room of a pueblo might accidentally catch fire, pueblo construction methods are not generally conducive to fire spreading rapidly from one room to another.

An unused or unroofed room in a pueblo is a different story, however. In the Southwest, where people often reuse the roof beams for new construction, unused rooms often become partly filled with organic as well as inorganic roofing debris, aeolian sand, and household discards, as well as any branches, brush, or tumbleweed that the wind blew in. Any organic material would likely be dry, easily ignited, and highly flammable. A lightning strike or a flying spark might easily spread to an abandoned room. This kind of fire is most likely to leave traces of heat damage on the room walls. In contrast, soot damage in rooms generally results from fire that lacks abundant oxygen, usually within an enclosed space (Adams 2004). In some archaeological examples, a relatively sharp line between burned and unburned masonry on a standing wall marks the height of long-gone piles of windblown sand that protected the wall base from the flame.

In this particular room with the unfired pottery, the upper walls would likely have experienced the most heat damage. There is evidence of some degree of burning on the sandstone blocks that tumbled down onto the floor, possibly from the upper portion of the walls. The burning, however, was not extensive enough to fire-harden the unfinished vessels. The adobe floor was not particularly reddened or baked from the fire, and no roof beams or extraordinary amount of charcoal was present in the fill. All this evidence suggests that there was a fire and people had to leave in a hurry, but the fire was self-limiting. The evidence suggests that people returned after the fire, cleaned up the room, made repairs, and plastered a fresh floor on top of some of the debris. The forty centimeters of debris left on the original floor seems to indicate that the inhabitants did not remove all of the fire debris. It is possible that some time elapsed between the fire and the time when this room was refurbished and reoccupied—at any rate, it is apparent that people did not feel the need to uncover or reuse the original floor surface.

Additional evidence of fire is found in Roomblock II. Here, the evidence points to more than one episode of burning. The fire-reddened portions of the east wall of Room 4 show evidence of intense heat in a limited area in the lower areas of the northeast corner, but the floor itself was not burned. This evidence is consistent with a fire igniting loose brush and debris inside an already abandoned room, where some midden or windblown sand was already present to protect the floor from contact with the heat. A second discrete level of fire-reddened masonry near the top of the preserved walls shows that a second period of burning occurred after the room filled with midden and sand.

In contrast, in another room (Room 3), the lower portion of the east wall shows evidence of more extensive burning, including fire-reddened rocks and adobe mortar, and also masonry that is black with thick soot. The room fill, however, represents the expected organic refuse and debris that one would expect in a midden context. The remains in Room 3 are more consistent with an episode of fire occurring during or just after the time of occupation.

In sum, the various rooms that show fire damage at Pueblo de la Mesa indicate that in some cases, episodes of burning occurred during the time when a room was occupied, or very soon thereafter. In other cases, episodes of fire post-dated people's active use of particular rooms or room blocks. In at least one case, people returned after the fire, cleaned up the debris, and reoccupied the same room. While this evidence provides a general sequence of events in certain individual rooms, we still do not know the length of time represented between them, or how significant these fire events were in the overall life of the community.

This information helps us understand the everyday activities of the pueblo's inhabitants and the general history of occupation of the pueblo site itself. The following sections examine how this information helps us understand the organization and operation of this pueblo as a village community.

Community Organization at Pueblo de la Mesa

In this section, I consider the available evidence for social groups and community dynamics at this Glaze A pueblo village site in more detail and evidence of possible changes over the constructional history of the site (table 17). While most of the evidence comes from architectural features, information from other datasets such as stratigraphy and artifact analysis also contributes to this understanding of the pueblo as a living community.

Evidence of Residential Groups

At Kite Pueblo (see chapter 4), I noted that the basic unit of repetition within the built environment includes two spatial scales: the individual room and the room block. At Pueblo de la Mesa, this repetition of rooms and room blocks is not so tightly organized, but shows many of the same social concerns.

A high degree of formal spatial patterning is expressed in the preplanned, ladder construction of the sandstone-walled Roomblock II. First, the rooms are virtually identical in size, suggesting a basic equivalency of units within the room block—an equivalency of size that was important enough to supersede any consideration of potential differences in room function between, for example, habitation room or storage room. The spatial layout of the room block was also important enough to supersede any smaller-scale social differences in, for instance, the size or composition of constituent subgroups such as nuclear families. The masonry construction at Pueblo de la Mesa makes it easier to establish that the room

Table 17. The history of Pueblo de la Mesa, LA 2091

Sequence	*Evidence/Location*	*Interpretation*
First occupation	Roomblock I area	A single curved line of rooms apparently lies under the surface architecture of Roomblock I. Excavation of Room 5 shows that room size was larger during this earliest occupation.
First construction of visible surface architecture	Roomblocks I and II, North Plaza	These two room blocks show planned (ladder) construction and define the North Plaza. There is a kiva (or at least a roofed semisubterranean structure) and a possible cistern on the North Plaza.
Remodeling and expansion	Addition to south side of Roomblock II (Room 4)	This addition expanded the living space of the pueblo.
Second plaza defined	Addition of Roomblocks III and IV	Additional room blocks enclosed and defined the South Plaza. Thick midden accumulated to the west of the site. At some point, people hastily abandoned the pueblo, at least temporarily, leaving clay pots drying in Room 3.
	South Plaza	The features on the North Plaza fell into disuse and filled with midden. The focus of occupation shifted to rooms surrounding the South Plaza. Some rooms in Roomblocks I and II were used for burials; other rooms filled with midden sediments.
Abandonment	Roomblock IV	This room block was the last to be occupied. Windblown sand accumulated in the empty rooms after abandonment (Room 8). There is evidence of fire damage in Roomblock II, but it could be post-occupational in date.

block was constructed in a given episode, according to a known plan, and was intended to accommodate a group of known size. The size and unified action recorded in this room block points to a relatively large and cohesive social group, with well-recognized and well-accepted boundaries. In addition, it is clear that this group was already in existence and that the members were already well acquainted with one another. They were able to predict what space would be needed to plan accordingly and to execute that plan fairly rapidly and with minimal internal variation in construction.

Based on the ethnographic record of historic Pueblo Indian groups, Southwestern archaeologists commonly identify such groups as multihousehold extended family groups such as a unilineal descent group, whether a single lineage or a portion of a larger clan. In many cases, such a group includes social investments in cooperative labor as well as co-residence. Although these ethnographic examples may not be strictly analogous to social groups that existed in earlier times, it is apparent that at Pueblo de la Mesa construction of Roomblock II testifies to the presence of at least one well-defined and bounded social group within the village site.

We also know that within this social grouping, differences in individual household or nuclear family size, wealth, or status were minimized—or at least it is clear that any such differences were not expressed in the built environment. The high degree of standardization between rooms testifies to the social and symbolic value that people placed upon reducing the apparent importance of any potential internal group differences. This strategy emphasized the importance of the group of people using the room block over any constituent subgroups, despite expectations that any such groups could be expected to vary in size and social composition at any given time and also over a number of years.

This basic spatial organization of Roomblock II as a double line of rooms (two rooms wide and six or eight rooms long) recalls the highly standardized room sizes and organization at the earlier site of Kite Pueblo, and also most likely present at the adobe portion of Frank's Pueblo. If indeed this architectural unit records some aspect of intrapueblo social grouping dynamics, it is obvious that only one such grouping was represented by the original construction at Pueblo de la Mesa.

Why only one such room block? At Kite Pueblo, the standardized room represents one level of social iteration and the standardized "ice cube tray" linear room block represents a second iterative element, forming the plaza-oriented pueblo as a whole. At Kite Pueblo, the similar size and shape of rooms and also the room blocks around the plaza de-emphasize social differences between room block groupings and accentuate the equality and standardized character of the pueblo's social sub-units. In this way, the iteration of spatially equivalent units of the built environment emphasizes the unity of the pueblo as a whole.

At Pueblo de la Mesa, a different social logic is apparent. While the original construction of Roomblock II indicates a similar concern for manipulating the built environment according to specific and standardized criteria, the variable size and organization of other room blocks within the pueblo and the sequential construction of the pueblo as a whole indicates that this particular logic was not sustained over time or across larger social groupings. Understanding the pueblo as a whole and the nature of the pueblo community that it represents thus involves more investigation into the ways that internal social differentiation was codified, expressed, and maintained in the built environment.

Evidence of Social Differentiation

At Pueblo de la Mesa, simple association or spatial mapping between a bounded kin group with a single architectural room block does not provide a very good model for understanding the pueblo as a whole. Because pueblo construction occurred over some period of time, however, we are able to get a better understanding of the social dynamics that were involved in new construction and also remodeling.

The earliest occupation of the site occurred in Roomblocks I and II. Continuing investment in construction and reconstruction at the village site indicates incremental growth of the village over time. It is not clear exactly how much time elapsed before remodeling began at the site—that is, whether we are considering a time frame of years or decades—but it is clear that subsequent construction indicates a widespread and enduring shift in the social importance and cohesion of these social groups. The second period of construction at the pueblo, characterized by the use of blocky limestone masonry, gives an overall impression of population growth by accretion. Incoming groups were apparently of different and variable size, and demonstrate differing relationships to the existing population.

Subsequent population growth was accommodated through two strategies, but these strategies may or may not have been temporally distinct. The first strategy distinguished here, which was not necessarily the first to occur, resulted in the construction of separate room blocks close by, but not connected to, the original building. This spatial separation suggests some acknowledgment of differentiation between social groups and possibly a degree of autonomy between residential groupings. A second strategy involved adding rooms attached to the southwest side of Roomblock II. The similarities and differences between these two strategies and their social significance are detailed below.

Both construction strategies were involved in the growth of the pueblo over time. It is interesting, however, to note also the care with which people minimized social difference across the site. First, room size is relatively constant across the site; each excavated room measured roughly two meters by three meters in size, which is a fairly typical room size for most of the Glaze A pueblos in Salinas. Thus, there seems to be no particular advantages that might ensue for earlier residents, nor do later migrants seem to be obviously advantaged or disadvantaged in terms of their access to basic living spaces.

A second type of evidence points to another social effort to minimize differentiation between the older parts of the pueblo and the newer parts. For example, in Roomblock II, the new limestone rooms along the southwest side of the room block are virtually identical to the old rooms in size and spacing, with each wall abutting against the older room block so that each new cross wall matches neatly to the cross wall on the other side. Whatever the purpose of these new rooms, for housing people or stockpiling foodstuffs, the inhabitants took care to follow the basic divisions of interior space that were already established. Thus, while Roomblock II may represent a relatively large and growing residential group, there is no obvious distinction in basic housing between the more established residents of that group and its newer members.

It seems clear that the new addition was carefully planned and executed with some effort to maintain established criteria, despite the difference in building materials. The fact that building material was apparently of lesser importance may imply that the whole construction was plastered over with adobe—if so, the room block as a whole would have presented a seamlessly unified and coherent appearance, despite the differences in the type of stone used for construction.

The most extensive additions to the pueblo, however, involved establishing new room blocks of variable size and shape, and with variable numbers of rooms. "New" construction may have been sequential in time, but adhered to a general overall plan that included spatially detached and defined room blocks organized around a central plaza. Room sizes and the overall arrangement of rooms into a contiguous room block, two or three rooms wide and of variable length, shows significant continuity over time in the conception and use of domestic space.

Overall, the architectural evidence suggests that while the original inhabitants did not necessarily plan for or specifically reserve areas for predicted growth in this locale, the periods of construction that did occur exhibit advanced planning regarding the desired spatial arrangement of room blocks within the site as a whole, and foreknowledge of the size of each room block as it proved to be needed. In addition, the new room blocks do show clear coordination with previous construction.

The "new" room blocks show that groups shared many ideas about proper room size, orientation, and construction with the previous inhabitants. Residents and newcomers also cooperated enough that they were able to define and maintain the open space of the lower plaza over time, suggesting a high degree of shared knowledge and expectations for village organization and layout, and also a high degree of consensus regarding the importance of this feature within the village site. In sum, the archaeological evidence provides an overall picture of a resident population and pueblo growth through the addition of spatially separated but highly similar room blocks.

This pattern of growth suggests some sort of social change involving discrete groups, but we do not know the details about the origins of the people who would have lived in the new room blocks. Do these new groups represent intravillage population growth over time? If so, people did not expand the living space as needed by incremental room additions, but rather made the decision to delay adding any rooms to existing room blocks until they were able to establish a whole new room block. It is also possible the group occupying the new room blocks originated from outside the village site. In either case, the similarity of architectural design and the placement of the new room blocks indicate that the groups shared many common ideas of architectural size, organization, and orientation of structures within the overall built environment. This shared patterned behavior suggests that the social distance between the inhabitants of the older and newer room blocks was relatively low. It also reinforces the idea that for the village community as a whole, the concept of the room block grouping was of paramount important in people's experience of the built environment.

Evidence of Social Integration

Aspects of group social integration include the development of shared, public space such as one might see in open plazas or semisubterranean kivas (e.g., Lipe and Hegmon 1989). These kinds of structures may represent different dimensions of social integration, however, in that activities on the plazas are typically highly visible to all members of the village site and any visitors, while most kivas represent smaller, interior space where activities would not necessarily be visible to the casual passerby.

The very large great kivas and also large unroofed kivas known across much of the American Southwest represent a later development, often interpreted as a scaling-up of kiva ritual to include a much larger audience.

These large kiva structures are common at Chaco Canyon, of course, but are also recorded in the Cibola area (e.g., Kintigh 1994) and near the Mesa Verde area (Adler 1994). Some of these large, circular, and roofless enclosures have been compared to a modern amphitheater (e.g., Kantner 2004:151).

The very large kiva structures such as those recorded in the Cibola area, measuring thirty-three meters or so in diameter (Kintigh 1994), are not known in Salinas, even among the later sites. However, smaller kivas measuring some two to four meters across are present. I have noted before (see chapter 4) that circular depressions at many different early pueblo sites in the Salinas region may or may not actually represent ceremonial chambers. In hindsight, I suspect that some of these circular depressions are most likely cisterns or water storage features. Some, such as at Kite Pueblo, are located in the interior plaza; at other sites (e.g., LA 9016 on Turkey Ridge), an obvious cistern is located a short distance away from the pueblo. There is no doubt, however, that kiva structures were very much a part of the built environment in the later pueblo sites, and in fact continued to be used well into the Spanish occupation (Wilson 1993).

The kind of social integration that is expressed in kiva architecture appears to be an important component of many early pueblo sites across the Southwest. On the Colorado Plateau, so-called proto-kivas date to about A.D. 700 or 800, and are associated with the first use of aboveground habitation structures, the so-called unit pueblo (see Lekson 1988).

The appearance of ceremonial architecture such as kivas and the ratio of kivas to domestic rooms are often considered to be indicators of the scale of ritual activity within a village site. For example, at Sand Canyon Pueblo in the northern Southwest, nearly every household is associated with a kiva structure (Adler 1994:97). Presumably, the locus for most ritual performance was in the residential household itself. In contrast, in some areas of the Pueblo Indian Southwest, larger social groupings associated with kivas are seen in the late A.D. 1200s to early A.D. 1300s. This shift is manifested in a larger ratio of domestic rooms (residential and storage rooms) per kiva structure (Steward 1937; Dean 1996). Adams (2002), among others, argues that this shift suggests that lines of cooperation and alliance between related households were increasingly formalized, with different lineages coalescing into larger bounded units such as clans.

At Pueblo de la Mesa, it is clear that a formal semisubterranean kiva was a significant architectural feature of the North Plaza. The kiva measured about 4 meters in diameter, and was dug about 1.4 meters into the limestone bedrock. An encircling limestone wall increased the interior height to at least 2.2 meters. Two postholes (18 centimeters in diameter) found near the perimeter of the structure indicate that the structure was roofed. The exposed interior of the limestone walls was finished with adobe and a thin layer of pink plaster.

The location of this ceremonial structure in the North Plaza suggests that it was an important part of the village community in the earlier phase of site occupation, when Roomblocks I and II were occupied. At this time, the North Plaza would have functioned as the central plaza, and the kiva would have been a prominent feature on the plaza.

The other structure on the North Plaza, originally identified as a kiva ("Kiva 2"), showed no sign of roofing or other alterations. In retrospect, I interpret this structure as a cistern or water storage feature. Midden deposits within both structures on the North Plaza indicate that their uses changed throughout the site occupation. At some point, both structures were repurposed for debris disposal, including ceramics, lithics, animal bone, and other discards of daily life. It seems likely that this shift in structure function may have occurred at the time when the South Plaza was defined, and many social, political, and economic activities shifted to that area.

This plaza is considerably larger, measuring about twenty-four to twenty-eight meters in diameter. In contrast, the smaller North Plaza was about half that size, measuring twelve meters between Roomblocks I and II. Given the space occupied by the two plaza features, the open space on the North Plaza would have been severely restricted. One could argue that this shift may represent a change in the use of the plaza over time for ceremonial gatherings. On the other hand, the larger South Plaza is comparable in size with the plaza at Kite Pueblo, which presumably dates earlier in time. It is apparent that some other social factors are impacting plaza size. For example, given that some early jacal sites were organized with clear and level exterior space outside and adjacent to the village site itself, it is entirely possible that the area of the South Plaza at Pueblo de la Mesa was reserved for organized group activities long before it was enclosed by later residential construction.

The closed nature of the South Plaza, however, was obviously of some importance, so much so that the pueblo's inhabitants went to some effort to close off and define the plaza deliberately. While at Kite Pueblo the enclosed plaza is integral to the entire site layout, at Pueblo de la Mesa people filled in the space that remained between spatially separated room blocks with a substantial masonry wall. At the time of excavations in 1992–1993, this wall stood shin-high. It is possible that the original height was about knee-high (about fifty centimeters), but not much higher. The rocks are much larger than the usual masonry blocks, and there is no evidence that the rocks were shaped or set with mortar, or even that they were placed very carefully adjacent to one another. In this respect, they appear less like a proper wall and more like a line of large rocks. The rocks were located several meters away from the drop-off at the edge of the mesa, and could hardly have been intended to prevent access or fortify the pueblo in any significant way. Instead, I have argued (Rautman 1995c) that the wall is better interpreted as a social symbol, defining the social space of the enclosed plaza and linking established portions of the village site (Roomblock II) with later construction (Roomblock IV).

It is possible, too, that the site's inhabitants were in fact planning for future migrants who would need additional housing and planned the spacing between different room blocks deliberately, reserving that southeastern portion of the plaza for an anticipated later arrival. In this respect, the plaza wall may have acted as a placeholder, or perhaps ultimately a recognition that the expected group was not going to arrive after all. In a literal sense, the plaza wall does close off the open plaza at the site, but this act may have had other significance in defining the boundaries of the pueblo community or expressing more actively the concept that the pueblo was now "closed" to outsiders.

In sum, the size of the South Plaza argues for the importance of site activities in an area large enough to include most if not all of the pueblo's residents. Intrasite social groupings are expressed in the form of different room blocks, and their sequential construction suggests a process in which discrete groups of people moved into the pueblo. There is also some social effort expended to accommodate the new arrivals within a familiar spatial form of the plaza-oriented pueblo.

The earliest construction on the site may also indicate social strategies toward whole-group integration. The earliest room block, Roomblock I, is somewhat different from the rest of the room blocks. It is a single line of rooms, rebuilt from an earlier buried structure that originally had rooms somewhat larger in size. The surface architecture of Roomblock I is also somewhat different from the usual Salinas room block. Unlike the straight linear room blocks seen on Jacal period sites and at Kite Pueblo, these contiguous rooms form a gentle curve, and the kiva is located at about the middle of the shallow concave arc of rooms. There is no way to establish whether Roomblock I and the kiva represent a single occupation, but the spatial organization suggests that such an association is possible.

Regardless of the exact chronological relationship of these structures, it is apparent that the restricted type of ritual that occurred inside kivas represents a different kind of social integration than that enacted within, and represented by, the open plaza. Future archaeologists may yet find more buried kiva structures on the South Plaza—right now we simply do not know if there are any. Regardless, the data that are available now argue that the residents of Pueblo de la Mesa, like Kite Pueblo, were quite concerned with establishing and reinforcing their shared identity and common interests, and that they expressed this concern through the form and construction of the built environment. Newly arriving groups maintained some degree of differentiation, however, at least in the form of discrete and spatially separated room block units. The organization of the site as a whole, however, communicates a continued and active creation of pueblo-wide organizational structure, including the concept of the central plaza.

Evidence of Social Commitment to Place

In the area around Chupadera Mesa, there is significant evidence of social groups maintaining a long-term commitment to specific locales. In the western portion of Salinas, for example, each known masonry pueblo that is classified as a "Glaze A pueblo" shows signs of an earlier Jacal period occupation.

Around Jumanes Mesa, however, no Jacal period sites have ever been identified (Caperton 1981:4). There are in fact four pithouse sites mapped in this area around the East Mesa, so it is possible that the absence of jacal sites does record local abandonment of this particular area. If so, the three early masonry pueblos of East Mesa—Pueblo de la Mesa, Pueblo Colorado, and Pueblo Blanco—represent a substantial population movement beginning around A.D. 1300 and coinciding with the widespread adoption of Glaze A pottery. Mera (1940) suggested long ago that all three of these sites were defensive in nature; Stuart and Gauthier (1981:323) take this argument a step further in suggesting that these upland sites across the Salinas region, and also in the Puerco-Salado region farther west, were frontier settlements representing the southernmost extent of the Pueblo Indian occupation of the Southwest. In Salinas generally, the concept of expanding frontiers and new construction of masonry towns does not fit with the archaeological evidence that suggests long-term and continuous occupation as well as a highly localized process of aggregation from jacal to pueblo sites.

The frontier concept fits better, however, within this eastern subregion of the Salinas area—at least as far as Pueblo de la Mesa is concerned. The nearby sites of Pueblo Colorado and Pueblo Blanco, however, are located in sandy lowland areas, and show signs of occupation during the full range of the Rio Grande glaze sequence. We do not know the details of construction and settlement history for these larger pueblos, however. To the extent that this cluster of pueblos along Jumanes Mesa does represent an expansion of the Pueblo Indian occupational frontier, however, this strategy seems to have been relatively long-lived and successful.

Glaze A Masonry Pueblos and Communities in Salinas

Looking back at this evidence, we can evaluate the ways in which this particular pueblo village site compares to our general understanding of a village community. The available evidence from Pueblo de la Mesa argues for an interpretation of this village site as a different kind of pueblo than Kite Pueblo. Whereas Kite Pueblo records one major coordinated building episode with relatively minor later alterations and renovations, the archaeological site of Pueblo de la Mesa records a more complicated history.

At the earlier site of Kite Pueblo, we have an impression of a single social group whose members shared a conception of overall village layout, who were already used to working together, and who spent considerable social energy in the design and construction of a village site. The built environment downplayed the significance of any existing social differences or village subgroups and instead emphasized similarity and interdependence at every social scale, including similarity of rooms, of room block size and layout, and of the village site as a whole. Later modifications involved expanding habitable space, but with considerable care to retain the original overall plan.

In contrast, Pueblo de la Mesa gives the overall impression of a village site constructed from separate "building blocks" at the social scale represented by the room block grouping. If one assumes that this group represents a lineage or extended family group, then Pueblo de la Mesa records a process of local aggregation of such social units. Unlike in the western portion of the Salinas region, there are no known earlier dispersed jacal village sites nearby, so indeed it is possible that people did arrive from some distance away, perhaps coming from more than one specific locale.

The earliest inhabitants of Pueblo de la Mesa did not live in a tightly organized plaza-oriented pueblo. Rather, their village site, a single arched line of rooms, is suggested by the presence of large rooms that were identified underneath the surface architecture of Roomblock I (see table 17). This earliest line of rooms, fronting a shared open space and possibly a kiva, recalls (as Mera 1940 suggested) the organizational layout of the small unit pueblos of the northern Southwest. The rest of the time of site occupation, however, is different. This difference is expressed in a different size of rooms (smaller), and a different room block layout, including ladder construction and parallel lines of rooms. The later construction (Roomblock II) and subsequent additions (Roomblocks III and IV) show that these inhabitants shared many general expectations about the spatial organization of the village community, a plan that was maintained across several different phases of construction.

So was Pueblo de la Mesa a defensive site, a fortress? Although some archaeologists might argue that plaza-oriented pueblos are defensive by definition, the earliest plaza-oriented pueblos in the Salinas area show little sign of any other defensive construction. After all, the adobe Kite Pueblo is just sitting out there in the Medanos Plain at the base of Chupadera Mesa. On the other hand, of course, the internal plaza and cistern would provide some facilities for people to retreat to the safety of the pueblo's interior, regroup, and possibly outlast their attackers, at least for a while.

In contrast, the two "black-on-white" pueblos that I mentioned at the beginning of this chapter seem far more likely candidates for defensive sites. I have not visited the one near Abó, but I do know from personal experience that in the southern portion of Salinas, Pueblo Seco (LA 9029) is located on a promontory so steep that it requires scrambling with hands and feet to reach it. I will discuss them further in the next chapter, in the context of regional relationships among the pueblos.

Pueblo de la Mesa, on the other hand, seems to be reasonably typical of the Salinas Glaze A masonry pueblos, at least in terms of its elevated location. It is located on an upland, possibly defensible setting, with an excellent view of the lowlands, but the mesa itself was not so very steep or inaccessible. The low wall defining the South Plaza could indeed be a defensive structure (LeBlanc 1999:56).* If so, it might be useful to slow down an intruder or to provide cover for an archer lying behind it, picking off enemies as they charged or sneaked up the last couple of feet to the top

* It is possible, however, that the low wall would have supplied some cover for an archer who might crouch behind it. Short curved walls of similar height, with the concave side facing the pueblo, can also be seen north of Pueblo Seco. This north side is the only portion of Pueblo Seco that is not protected by steep rocky slopes and provides the only easily travelled route of access into the pueblo.

of the mesa, scrambling over the sharp edges of bedrock and limestone boulders. On the other hand, this knee-high construction might just as well represent a social boundary marker.

For example, at Pecos Pueblo, a low encircling wall was originally assumed to be a defensive wall (Kidder 1924:66), but more recent descriptions argue for its role in social marking—"inside" versus "outside" the pueblo (Kessell 1979). During the Spanish period at Pecos, non-residents such as traders from the Plains, in this case Comanche Indians, would traditionally camp outside the wall; by custom, they were expected to leave the pueblo by nightfall.

Other archaeologists have questioned the widespread assumption that upland pueblos are necessarily defensive sites. One idea is that highly visible upland pueblos may be important in social signaling and territorial marking. Some of this signaling may be explicitly militaristic in nature, but other aspects of territorial marking also involve more subtle messages. For example, Altschul (1978) makes this argument for the Chaco area. Adler (1994, 1996) has made a similar point regarding the archaeology of Crow Canyon and Mesa Verde, and Kohler (1992) also relates the location of field houses to advertise use rights to land and resources. There is actually quite a lot of writing on this concept of social signaling (Roscoe 2009). In its more extreme forms, it is called wasteful advertising (e.g., Aranyosi 1999; Dunnell 1999) and is relatively common in ancient as well as modern societies. These studies consider the trade-offs involved in what we might interpret as useless investment in monument construction or other forms of what Thorstein Veblen (1899) called conspicuous consumption. They address particularly how such actions may not be that wasteful or misguided in certain contexts and attempt to articulate the social and environmental conditions under which such a strategy might make sense (Madsen et al. 1999).

Stone and Downum's (1999) comparative study of aggregation in the ethnographic and archaeological record is particularly interesting in this respect because they articulate a specific social mechanism by which archaeological variables such as site size and location translate into the social idea of rights to land. They argue that large and highly visible pueblos not only physically mark territory but also provide a material symbol of a group: the size of the pueblo demonstrates to outsiders the size and cohesiveness of group membership, even if not all group members live at the site. From this perspective, large pueblo sites need not be defensive per se and were not necessarily designed for actual defense in warfare. The general idea here is that any group that is sufficiently large, well organized, and unified enough to construct a visually imposing pueblo would also be able to recruit members to back up their territorial claims with force if necessary (see also Tringham 1972:470; Leone 1987; Lowell 1996). This advertising strategy might therefore be the complete opposite of wasteful in certain contexts. Apparently, an analogous situation also occurs in some medieval English fortifications and castles—other factors besides defensive functions were significant in structuring the use of space and the design of their construction (Kenyon 1990). But they look enough like how castles are supposed to look to convey the intended message of defensive potential, while also communicating other political and social meanings to the observer (Johnson 1999:156).

Traditional public dances at modern pueblos may be in some respects a comparable but more complex exercise. That is, in addition to the other purposes or functions that the dance may involve, the sheer fact of the performance may also communicate other social messages to viewers (e.g., Moore 1996; Low 2000; Inomata 2006; Ruscavage-Barz and Bagwell 2006). According to this idea, public performances would represent a way in which a given pueblo demonstrates its strength in numbers to visitors and neighbors by virtue of the fact that so many of its members return home and participate in the dance, even if they may not reside full time in the pueblo itself. One can imagine a situation where neighboring groups might visit such dances, not only for the social and ritual reasons but also as a means of assessing the general situation—how many people the host pueblo is able to supply for the dance, the dancers' energy and dedication to maintaining proper costume, equipment, actions, their families' willingness and ability to supply food and other needed supplies, the number of children coming up in the ranks of potential warriors, and the number of non-resident members who are making the effort to show up for the event. These are all useful things to know about one's neighbors, regardless of whether one views them as potential enemies or potential allies.

In sum, there are several alternative ideas that might explain the widespread adoption of the plaza-oriented site layout across the Pueblo Indian Southwest, as well as the upland location of these Glaze A pueblos in Salinas. At Pueblo de la Mesa, there seemed to be no particular reason to infer that the site's inhabitants were necessarily concerned solely or even primarily with defense.

With these ideas in mind, I began archaeological research at another Glaze A pueblo, Frank's Pueblo. This pueblo is located up on the rim of Chupadera Mesa, overlooking Gran Quivira on the next little ridge to the northeast and the Gallinas Mountains in the far distance to the east.

Matthew Chamberlin's survey had established that in this area, each of the known Glaze A pueblos was situated in the midst of an earlier dispersed jacal community (Chamberlin 2001, 2008, 2012). I thought that Frank's Pueblo would make a good comparison case study for understanding the process of aggregation during the early pueblo period.

It took only a small amount of excavation at the site, however, to appreciate that all these ideas would recede in significance compared with the unexpectedly dramatic events that were recorded at the site in both architecture and in the cultural remains. While the excavations at Pueblo de la Mesa presented an image of ordinary daily village life, and a process of orderly expansion by internal growth or possibly immigration of (likely) socially and culturally similar groups, at Frank's Pueblo this rather cozy scenario was disrupted by our discovery of catastrophic widespread fires and evidence of repeated efforts to reoccupy and rebuild the site. In the next chapter, I consider this evidence from Frank's Pueblo in more detail, and discuss its implications not just for interpreting the individual archaeological sites described here but also for understanding their changing regional contexts.

CHAPTER SIX

Pueblo Communities and Regional Interactions

All defensive sites [in Salinas] came into existence during Group A times, or in other terms Period 1, none being founded at any later period.
—H. P. Mera, *Population Changes in the Rio Grande Glaze-Paint Area*

The early masonry pueblos such as Pueblo de la Mesa have been particularly interesting to archaeologists, partly because of their association with the first import of Rio Grande glaze pottery from the Galisteo Basin and partly because of the possibility that they did signal the beginnings of organized intergroup conflict. The quote from H. P. Mera above, based on his fieldwork from the 1930s, thus represented archaeological common knowledge for over seventy years. In fact, until relatively recently, no one knew whether all these apparently new Glaze A masonry pueblos in upland settings really represented a shift in settlement pattern or whether people had been living there all along in the jacal structures that are not so visible in the archaeological record.

Based on Matthew Chamberlin's survey, however, we now know that Jacal period village sites surround each of the masonry pueblos near Chupadera Mesa, thus showing that both upland and lowland regions had long histories of occupation (Chamberlin 2006, 2008). Our joint fieldwork at Frank's Pueblo also established that an early adobe pueblo lies buried underneath the visible masonry of at least one of these pueblos and possibly others as well. These lines of evidence argue for a great deal of continuity in people's occupation of specific locales, without any abrupt shift in settlement locations for the Glaze A pueblos.

In contrast, the occupants of the Jumanes Mesa sites on the eastern margins of the Salinas region seem to have experienced a different local history, with a hiatus of occupation during the Jacal period. In the 1930s, Mera suggested that the Jumanes Mesa sites represented frontier outposts on the edge of the Pueblo Indian world (Mera 1935). This idea of frontier settlements may in fact have some validity for the Jumanes Mesa sites, at least in terms of conveying the idea of a relatively small initial founding population who came to the area during Glaze A times, establishing pueblos such as Pueblo de la Mesa (as well as Pueblo Colorado and Pueblo Blanco) in previously unoccupied areas.

Was this "frontier" experience also applicable for other early masonry sites in Salinas? Since nearly all of the existing information about the Early Pueblo period was derived from the very large and complex Spanish mission sites, no one really knew much about the Glaze A occupations of the other Salinas pueblos. With this problem in mind, I began a research project at Frank's Pueblo (LA 9032) on the eastern slope of Chupadera Mesa during 2007 and 2008, assisted by Matthew Chamberlin (Rautman and Chamberlin 2008; Chamberlin and Rautman 2009; Chamberlin et al. 2011). Information from this pueblo would provide a way to compare the Jumanes Mesa sites (Pueblo de la Mesa) with a pueblo of comparable time period from around Chupadera Mesa. And, in fact, the history of construction, use, and abandonment of this site turned out to be quite different.

In this chapter, I summarize the known variation among the Early Pueblo period sites within the Salinas region, the inferences that scholars have made about their defensive functions, and I describe my recent research at Frank's Pueblo in light of this background. As it turns out, the evidence from this site did not just contribute to our understanding of this particular village site and its local community dynamics; it also raised significant and unexpected questions regarding regional social and political relationships. I supplement the information from Frank's Pueblo with published information from other sites, with personal observations, and also some information from recent excavations by the James Madison University (Virginia) archaeological field school, directed by Julie Solometo and Matthew Chamberlin.

Regional Perspectives Regarding Early Pueblo Communities

There are a number of Early Pueblo period sites in the Salinas region, but only a few of these pueblos continue to be occupied until Spanish conquest (see the list in table 13). In the southern portion of Salinas, most of the Glaze A pueblos are abandoned before about A.D. 1425, when

Glaze B pottery began to be manufactured and imported into Salinas; only Pueblo Pardo and Gran Quivira continued to be occupied after that time. In the northern portion of the region, only two sites, Abó and its smaller neighbour Ténabo, represent the late period occupation; in the eastern portion of the Jumanos region, Pueblo Blanco (Tabirá) and nearby Pueblo Colorado continued to be occupied.*

The Tiwa-speaking pueblos of Quaraí, Chililí, and Tajique, located along the east side of the Manzanos Mountains, were occupied through the Spanish period; in fact, Chililí and Tajique are still part of the old Spanish land grants in New Mexico, although they are no longer Indian pueblos. However, these pueblos were not necessarily occupied continuously throughout the centuries, as recorded by the glaze paint pottery sequence (Tainter and Levine 1987).

Although the details of occupation of all the large sites are not known, we do know that Quaraí apparently lacks much Glaze C pottery, suggesting that the site was abandoned during the mid-1400s; the presence of abundant Glaze E pottery indicates reoccupation in the A.D. 1500s (Hurt 1990; Spielmann 1993, 1994a; see table 14 in this volume for the manufacturing dates of glaze ware types).† At Gran Quivira, Hayes (1981:2) found similar evidence of periods of site abandonment for relatively short periods of time during the late 1400s and early 1500s, with renewed and extensive construction beginning about A.D. 1545.

The important point is that although the large late pueblos grew in size after many of the small Glaze A masonry pueblos were abandoned, this process of growth was not necessarily continuous or linear. Unlike the situation in the Rio Abajo, where population growth shows dramatic increase over a short period of time, the growth of the late pueblos in Salinas likely

---

* Pueblo Pardo seems to have been abandoned soon after Spanish conquest; Toulouse and Stephenson (1960) use ceramic evidence to argue for an abandonment date of 1630. There is some evidence that the Spanish sometimes lumped nearby sites together, either literally by moving people from the smaller to the larger pueblo or administratively by using only the name of the larger pueblo. Gran Quivira and its smaller neighbor, Pueblo Pardo, likely experienced this melding of identities (Toulouse and Stephenson 1960), and a similar process may have affected Abó and its smaller neighbor Ténabo (Hayes et al. 1981). Pueblo Colorado likely was not occupied at the time of Spanish conquest (Hayes et al. 1981).

† A similar complex history of pueblo abandonment and reoccupation around this same time (the mid-1400s) seems to be common among the pueblos surveyed by Montgomery and Bowman (1989) in nearby Chupadero Arroyo (west of the Jumanos pueblos of Salinas). See also Murrell (2006) for a recent discussion of water management features around these pueblos.

incorporates some rate of natural increase and also some reassortment of the existing Salinas population from the several smaller pueblos into fewer nucleated sites.

Across the landscape of Salinas, we might imagine the small early village sites as locally dense population clusters, their inhabitants related to one another by at least some sense of shared assumptions and knowledge and some shared history as long-term occupants of the Salinas region. There were likely tacit and possibly explicit agreements regarding the innumerable possible uses for cultural and natural resources within the area, including, for example, agricultural land, construction timbers, firewood, water, and pottery clay. People left traces of their presence on the landscape through use over time, cutting down the trees for construction and firewood, digging shallow reservoirs and water storage features, and wearing trails between village sites and various resources.* Some portions of one well-worn "Indian trail" that was also used by the Spanish explorers and missionaries can be seen in the area north of Gran Quivira

---

* One way to estimate the general scale of the geographic area around a pueblo that was habitually used for firewood collecting, gardening, and gathering is by using the (probably minimal) area estimates used by the Spanish administrators for establishing encomiendas. According to Spanish law at the time, privately owned ranches or estancias could not encroach within fifteen kilometers of a pueblo, although in practice apparently some ranches were allowed within four kilometers of a pueblo. We know, for example, that there were six private ranches between Quaraí and Chililí, a distance of less than fifty kilometers (see discussion and citations in Ivey 1988:26–27).

Throughout this chapter and volume, the term "abandonment" is used in the archaeological sense of disruption of residential continuity. The concept of abandonment in the archaeological literature is a complex one; see Nelson and Hegmon (2001) and Nelson and Schachner (2002) for discussion. Empty rooms and structures are not necessarily neglected or rejected simply because their use has changed. Nor have the residents or their descendants necessarily forfeited rights of use of a seemingly abandoned structure.

A local Salinas example might be the historic Euro-American town of Gran Quivira, located near the pueblo of the same name. There, former residents and their families may still hold legal title to the land, hold reunions, visit, and value these locations; yet archaeologically speaking, it is clear that most buildings are abandoned in the sense that they are no longer occupied or actively maintained in the way they were in the past. Similarly, even though the families of the Gran Quivira community still exist, interact with one another, and so embody this social group, it is nevertheless true that the resident community at the historic site of Gran Quivira is no longer.

In some respects, the modern American concept of a ghost town does have some relevance to this discussion of abandonment, reminding us that a town has an existence apart from just its architecture or physical location, that its existence involves another world of meaning. One can interpret the concept of the ghost town in the sense of

today (Leroy Nix, National Park Service ranger, personal communication 2008). People also marked significant locales with shrines and rock cairns (Chamberlin 2006), and extensive panels of rock art. Petroglyphs in Abó Canyon and pictographs on sheltered sandstone outcrops near Abó Pass, Arroyo Seco, and also Pueblo Blanco also attest to the importance of these specific locales (Cole 1984; Schaafsma 1993).

Although in some respects we can consider the Glaze A pueblos as a group, there are significant differences between them. Here, I briefly review three issues related to the early masonry pueblos: first, the unusual circular pueblo at Gran Quivira, and second, the observed spatial clustering of the later pueblos within the Salinas region (e.g., Mera 1940). Third, I discuss the apparently defensive location of most early pueblos and describe how excavations at Frank's Pueblo contribute to our understanding of regional interaction and possible conflict among the pueblos during Glaze A times.

The Early Circular Pueblo at Gran Quivira

While the majority of early pueblos in Salinas generally exhibit a common layout with four linear room blocks surrounding a central plaza (or some variation thereof), the earliest masonry pueblo at Gran Quivira is quite different. It exhibits an unusual circular village plan with rooms surrounding a circular central plaza, the cross-walls arranged like spokes of a wheel from axle to rim. Sites with a curvilinear layout are relatively common in the northern Southwest, with most dating to the Pueblo III period, after A.D. 1200. In these sites, a line of contiguous rooms forms a broad arc as if arranged along the perimeter of a circle. In addition, of course, D-shaped pueblos such as those at Chaco Canyon also include an arc or semicircle of habitation rooms fronting on shared plaza space. In both village layouts, the cleared space in front of the arc of rooms commonly includes one or more pit structures or formal kivas.

However, pueblos arranged with habitation rooms forming a full circle are less well-known (see discussion in Morgan 1994). There are at least six known in the Zuni area, plus one site farther away, Kintiel, which is considered to be ancestral Zuni (Hayes 1981:20, and references cited therein; Spier 1917; Fewkes [1904]2009:125). Fire House, a pueblo north

---

a haunted house, one in which restless spirits maintain occupancy in some state of conflict with the living; on the other hand, another conception of the ghost town emphasizes a complex dynamic between past and present, including a sense of place, of belonging, and of identity that unites and also differentiates the dead and the living.

of Keams Canyon in Arizona, is another known circular pueblo dating to around the same time (Hayes 1981:20, citing Mindeleff [1891]1989). East of Zuni, circular pueblos include Tyuonyi in Bandelier National Monument (Kohler 2004) and the one at Gran Quivira (Hayes et al. 1981).*

Gran Quivira's buried circular pueblo was discovered during the National Park Service excavation of Mound 7. The walls were set directly on a limestone ridge that rises above the level of the Medanos Plains below. Hayes describes the pueblo as built within a relatively short time using a preconceived plan, forming a "wheel-like" house approximately 43.6 meters in diameter, with a central plaza measuring some 18 meters across. A pit interpreted as a kiva was located in the center of the circular plaza (Hayes 1981:20). The room block varied from four to six rooms deep (plaza side to outside), although it was possibly only two rooms deep on the northern third of the circle (Hayes 1981:20). He concluded that the concentric

---

*Ivey (1988:15, note 18) reports that the earliest investigator, Edgar L. Hewett, believed that the 1913 excavations revealed evidence of a circular pueblo under Mound A at Quaraí. Al Hayes (1981:20) and Walter (1931) also mention the possibility of this circular pueblo, but later testing by other investigators showed only rectangular structures (Ivey 1988:15, note 18). Ivey suggests that later test excavations may not have been deep enough to reach the earliest occupation levels. Based on her excavations at Quaraí during the 1990s, Spielmann (personal communication 2012) is more pessimistic about its existence at all.

At Gran Quivira, Hayes et al. 1981 suggested that there might be two circular pueblos: one under Mound 7, the focus of the 1960s excavations, and a second possible circular pueblo is located just to the south, under Mounds 7 and 8 (Ivey 1988:17). Excavations here in the 1960s showed only a south-facing arc of 24 rooms. Based on this information, Hayes defined a second circular pueblo, which he called Feature 1A (Hayes 1981:19ff). Hayes notes that the walls of Feature 1A are less substantial than in Feature 1, and that the room arrangements and wall abutments seem to be more variable and ad hoc than those seen in Feature 1. Based on this evidence, he suggests that Feature 1A was built rather hastily with looser adherence to a set overall plan. In terms of chronology, while both circular pueblos date to roughly the same time period, Hayes (1981:19ff) notes that all walls in Feature 1 are set on bedrock or culturally sterile clay, whereas a thin layer of midden sediment and ash was found underneath two walls of Feature 1A (Hayes 1981:20). He uses this evidence to argue that Feature 1 was likely built and occupied first. Aside from these few descriptions, however, not much is known about the circular pueblos in Salinas.

It is possible that this arc-shaped structure was in fact a large kiva; we simply do not know (Katherine A. Spielmann, personal communication 2012). An early rectangular room block at Gran Quivira, bordering the south side of the main plaza, may also date to this early time period. If so, it is possible that it was occupied at the same time as the circular pueblo, but again, we simply do not know enough about the early constructions at Gran Quivira to decide.

walls were constructed first, and were more substantial in thickness, and the radial walls were added later as partitions (Hayes 1981:15–20).

Hayes (1981) estimates that the circular pueblo contained a total of 160 to 209 rooms, depending on the exact number of complete circles (or tiers) of rooms. Rooms were roughly 4.6 to 5.6 square meters. Although the rooms were not strictly square since their walls were curved around the circular central plaza, the rooms were fairly similar to one another in size and uniformly trapezoid in outline (Hayes 1981:15–16). Room size varied from 1.3 meters to 2.6 meters with an average of about 2 meters across (Hayes 1981:15). The size represented here is comparable to other sites in Salinas, that is, a square room measuring some 2 meters to 2.5 meters on a side.

Were the people who built and occupied these circular pueblos any different from the people who lived in other early masonry pueblos? Did this new idea represent the influx of a "new" immigrant population from the west? In fact, there is little to no evidence that this unusual architectural form stems from any specific change in population movements, exchange networks, or other forms of social interaction. We know that Salinas groups were already importing Western Pueblo ceramic types such as St. John's Polychrome at this time and were therefore likely in contact with groups in that area, but there is no indication that this interaction had changed in any measurable way. In fact, Chamberlin's (2008) study of ceramic sources indicates considerable overall continuity in people's access to Western Pueblo ceramics (see also Rautman 2013b).

Vivian (1979:125) does argue for at least two episodes of population movement and immigration: an earlier one around A.D. 900 involving groups from the northern Rio Grande and evidenced by the development of surface houses and painted pottery in Salinas, and also a later one in the 1200s, as Jornada Mogollon groups moved from the mountainous regions of the Sierra Blancas into the northern Tularosa Basin and the Salinas region. Overall, however, Vivian (1979) seems to envision a scenario of relatively low-level population movements into and out of the Salinas region, as well as considerable continuity in interregional communication and social interaction, resulting in a Salinas population expressing an assortment of northern and southern traits. We also know from other research that Salinas populations and those in the Sierra Blanca region were communicating and trading with one another, as shown in the overall similarity of their ceramic assemblages (Rautman 1993b) and also by the high degree of similarity in design elements used on Chupadero Black-on-White pottery (Clark 2006).

In contrast, much later, extensive rebuilding at Gran Quivira that Hayes et al. (1981) dates to the A.D. 1550s is in fact associated with the introduction of new social practices. At that time, people at Gran Quivira began using a new funerary practice, cremating some but not all of their dead. At about the same time, people ceased production and use of Chupadero Black-on-White pottery with its geometric designs, changing to a new local pottery type, Tabirá Black-on-White, which emphasized more representational images (Vivian 1979).

These changes have been attributed to a possible influx of population from the west (especially from the Zuni area), but in fact there is little evidence of actual population movements (see Cordell 1995). McWilliams (1981:149) as well as Turner (1981:117), for example, found little evidence of biological population differences in dental or cranial morphology across time or even between the groups of cremated and interred skeletal remains at Gran Quivira. Turner (1981:120) explicitly considers and rejects the possibility of population in-migration from the Western Pueblo region, concluding that the Gran Quivira population more closely resembles that of people at Pecos Pueblo, another Plains border town (but see Morgan 2010 for a more recent reanalysis of the Pecos data). While there are some differences in cranial morphology over time, Reed (1981:109) attributes this difference to changes in cradleboard use. Recently, however, a more in-depth study of non-metric skeletal traits suggests that later populations at Gran Quivira did physically resemble those from the Zuni region (Peeples 2014), indicating that population influx from Western Pueblo areas may have been more significant than previously thought.

In sum, all this evidence points to a relatively stable regional population within Salinas, with some expected accommodations of both people and ideas from diverse areas over the centuries, but no obvious or dramatic influxes of large or discrete "outsider" groups comparable to the large-scale population increase that Marshall and Walt (1984) recorded for the Rio Abajo region. The idea of the circular pueblo during the Early Pueblo period, like the observed shift to cremation burials later in time, points to the arrival of new ideas and possibly new people coming from other areas, but these changes do not necessarily signify any particular sudden wholesale population immigration or replacement.

More generally, many archaeologists have tried to investigate the cultural significance of overall village layout in the American Southwest and other regions (see discussion in Rautman 2013b); various models argue explicitly for the association of village layout with other variables such as kinship groupings, clans, and lineages (e.g., Reed 1956), household organization (e.g., Flannery 1972, 2002), gender relations (e.g., Mobley-Tanaka 1997; Potter 2002), ideational or cosmological models (Ewart 2003; Means 2007; Psarra 2009; Bradley 2012), concepts of power (Nielson 1995; Graves and Van Keuren 2011), or cultural rules governing social interactions (Sanders 1990; Fisher 2009). Whatever meanings may have been conveyed by this circular pueblo at Gran Quivira, however, it is clear that the unusual layout marks this particular village site (and possibly its affiliated community) as different from the other village sites within the region, but we do not know what this special status may have entailed.

Early Pueblos and Pueblo Clusters in Salinas

Another consideration regarding regional relationships among the Salinas pueblos is their obvious spatial clustering (e.g., Mera 1940). For example, in the southern portion of Salinas, the large pueblo of Gran Quivira seems to be paired with the smaller nearby Pueblo Pardo. To the east, Pueblo Blanco is paired with Pueblo Colorado, and in the northern area, the pueblo of Abó is located close to a smaller pueblo called Ténabo (see figure 17). In each of these cases, one pueblo in each pair is apparently abandoned just before or after Spanish conquest, and occupation continues at the larger of the two pueblos. Among the Tiwa pueblos, Quaraí, Chililí, and Tajique form another loose cluster; Hayes (1981) suggests that Chililí and Tajique may have formed a similar pairing.

Spatially discrete clusters of multiple nucleated sites are known from other areas of the Southwest during the Pueblo IV period, for example, at Homol'ovi (Adams 2004) and Zuni (Duff 2002). Marshall and Walt (1984:135) note a somewhat similar phenomenon of paired pueblos among the Ancestral Piro pueblos, with the two pueblos of each pair located on either side of the Rio Grande. Pueblos in the Chupadero Arroyo survey area also seem to occur in pairs (Mongomery and Bowman 1989). Some archaeologists interpret such clusters as expressions of a political hierarchy (e.g., LeBlanc 1999). Others interpreted these clusters as an emergent form of the group identities that were later noted by the Spanish explorers and administrators (Duff 2002; Graves 2002, 2011; Huntley and Kintigh 2004; Bernardini 2005). In fact, there is apparently considerable variation in the internal social and political organization of these clusters. For example, some Zuni and Hopi pueblo clusters do seem to have

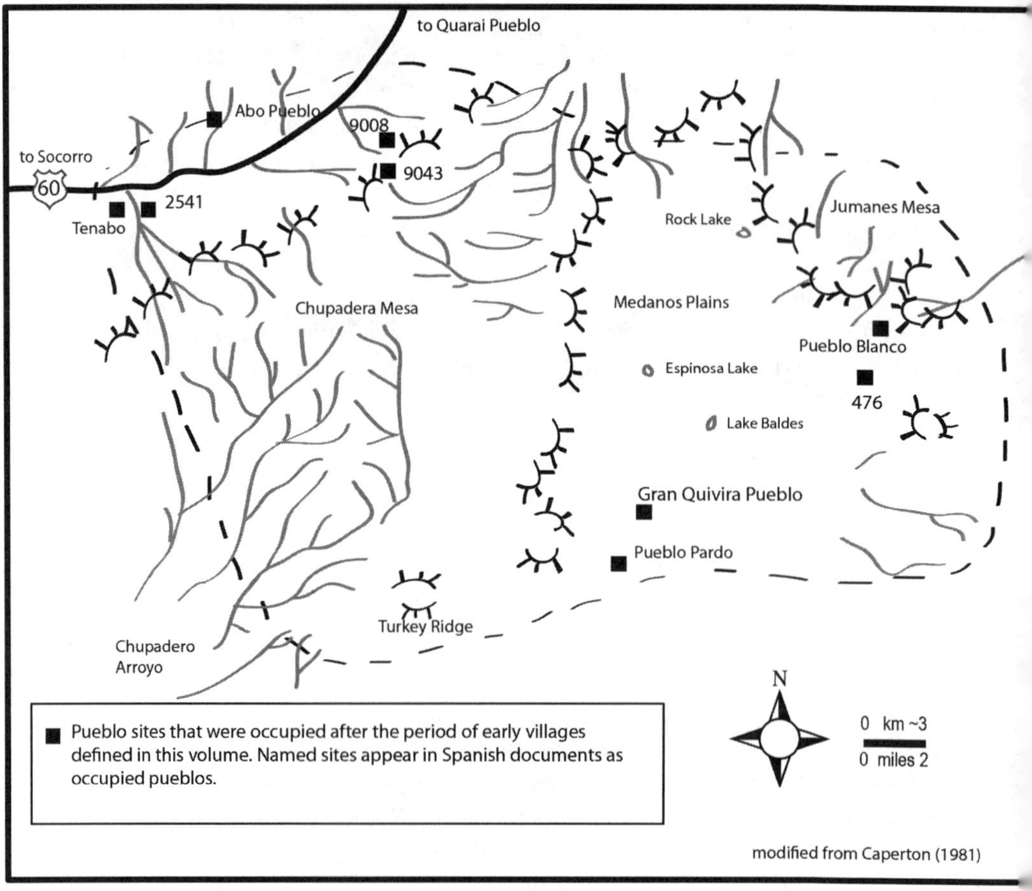

Figure 17. Location of late pueblo sites in the Salinas Province. Modified from the base map shown in Caperton 1981:xii.

been politically integrated (Spielmann 2004b). In other regions, member villages within each cluster may have retained considerable autonomy (Spielmann 1994b, 2004b; Adams 2002).

The four pueblos of Abó, Quaraí, Pueblo Blanco, and Gran Quivira formed a single spatial cluster that the Spanish later recognized when they grouped them together as the Salinas Province. We know that people in these pueblos were politically autonomous, but also highly interconnected in a complex set of social, economic, and political relationships. Spielmann (1994b, 2004b) has proposed a model that she calls a clustered confederacy to describe these relationships; Graves (2002) has utilized concepts of heterarchy (Brumfiel 1995) to develop more detailed models of regional interactions among the late Salinas pueblos (see also Graves and Spielmann 2000b). We know, for example, that the different pueblos had their own internal political organization, but that Gran Quivira seems to have played a major role regionally as the "first among equals," managing much of the local trade with the Plains Apache and hosting interpueblo feasts (see also Spielmann 1992, 2002, 2004a, 2011; see also Graves 2011).

What we do not know, however, is the time depth involved in these relationships. For the Glaze A pueblos, we do see some spatial clustering of smaller masonry pueblos around what later became the larger sites. In the eastern portion of the Salinas region, Pueblo de la Mesa is located near Pueblo Blanco and Pueblo Colorado, both of which grew in size after the period when Pueblo de la Mesa was abandoned. By the time of Spanish conquest in the sixteenth century, however, only Pueblo Blanco was occupied. In the southern portion of the Salinas Province, Frank's Pueblo (LA 9032) is one of several small masonry pueblos in the area around Gran Quivira (see figure 15). While most of these small pueblos were abandoned after Glaze A times, Gran Quivira, of course, continued to be occupied until and also after Spanish conquest. Certainly, by the end of Glaze A (the late A.D. 1300 or early 1400s), most of the small pueblos throughout the Salinas region are abandoned, and the large pueblos grow considerably in size. It is tempting to imagine that the large pueblos such as Abó, Gran Quivira, and Pueblo Blanco grew larger in size due to an influx of people from nearby sites. However, in fact, we really do not know whether this temporal relationship represents a true causal connection.

Overall, however, spatial clusters of early masonry pueblos show a similar occupational history across the different subregions of the Salinas Province. This situation would seem to make a perfect case study for comparing the Glaze A occupations of the eastern and southern subregions of Salinas, or so I anticipated when I planned my field project at Frank's Pueblo. In the end, however, these excavations provided the most information regarding defensive strategies and regional conflict.

Early Pueblos in Defensive Locations

As archaeologists have noted since the 1930s, another obvious characteristic of the early masonry plaza-oriented pueblos in Salinas is their upland settings (e.g., Mera 1940). Only LA 9021 is located in a less obvious position, on the rolling upland landscape of Chupadera Mesa, but it still commands a good view of its surroundings (Chamberlin 2008). Beers (2012) recently tested some of these ideas of interpueblo signalling among the Jumanos Division pueblos.

In general, however, many Southwestern archaeologists (for example, reviewed in Bernardini 1998) argue that the combination of upland setting and a closed-site layout is most likely an indication that the pueblo represents a defensive site (e.g., Stuart and Gauthier 1981; LeBlanc 1998, 1999). However, in my view, most of the upland Salinas pueblos are not nearly as inaccessible as some sites in other parts of the Southwest (e.g., LeBlanc 1999:fig. 2.5 and 2.6; see also Kenzle 1997). The two possible exceptions are Pueblo Seco near Gran Quivira and a site numbered LA 503 near Abó. Both are considered to be "black-on-white pueblos" (Montgomery and Bowman 1989), and apparently predate the use of Glaze A pottery. Because we know so little about LA 503, I will consider only recent research at Pueblo Seco below.

Most of what we know about Pueblo Seco comes from recent survey and small-scale test excavations by Julie Solometo and Matthew Chamberlin (personal communication 2013). The site is located at the top of a steep landform, at the top of a fairly abrupt rise on a mesa spur above Arroyo Seco. The pueblo itself is surrounded on three sides by a steep drop-off littered with fallen boulders and jagged rocks, accessible by scrambling with hands and feet up the steep mesa slope. Only a narrow neck connects the small mesa spur (and pueblo) to the larger mesa.

Pueblo Seco is also unusual in having no prior Jacal period occupation in the surrounding area (Chamberlin 2008). Solometo and Chamberlin's recent research confirms that this pueblo was constructed rapidly. It also exhibits several features that are defensive in nature, including a wall running across the promontory at its narrowest point, as well as several nearby freestanding masonry arcs that would provide shelter for an archer or patrol. In sum, despite the fact that many Salinas pueblos are located in upland settings, only at Pueblo Seco is there evidence to argue that defensive considerations were the initial and also the primary factors affecting site construction and use.

What no one knows, however, is the cultural context of this defensive positioning. The unanswered questions are seemingly endless: Why were such defensive constructions needed? Who was the enemy? Were the pueblo groups fighting one another, or were they concerned about attacks from "outsiders"? Pueblo Seco and also LA 503 near Abó Pass are located on the western border of the Salinas region; this location suggests that these sites may have played some sort of role in monitoring or defending a boundary between the Tompiro who occupied Salinas and the ancestral

Piro groups who lived to the west along the Rio Grande. Alternatively, these border positions could also represent a strategy of retreat from some threat originating farther east.

At Pueblo de la Mesa, these questions about regional political relationships were difficult to evaluate beyond noting that the evidence for everyday domestic activities, including ceramic manufacture, argued against interpreting the site as specialized fortress or refuge. However, there is no doubt that the upland location would have provided a distinct advantage to the site's inhabitants in monitoring the eastern borders of Salinas and the Plains beyond.

Excavations at Frank's Pueblo during 2007 and 2008 provided quite a bit more evidence of the details concerning regional conflict and its impact on the pueblo community. So with these rather lengthy introductory comments, I consider the evidence from Frank's Pueblo below.

Excavations at Frank's Pueblo

One major contribution of the research at Frank's Pueblo was confirming that people had lived in this particular locale for a very long time (Rautman 2008b, 2013a, 2013c). Chamberlin (2008) had already noted a Jacal period occupation nearby the mounds of jumbled masonry that mark the pueblo site, but our joint archaeological field school excavations showed that there was also an early adobe pueblo underneath the known masonry pueblo (figure 18). This adobe pueblo, like Kite Pueblo, consisted of four room blocks surrounding a rectangular central plaza.

The Glaze A Masonry Pueblo

Unlike Kite Pueblo, where occupation apparently ceased soon after the period of remodeling with masonry, the adobe structure at Frank's Pueblo continued to be occupied even after it was extensively modified and expanded by subsequent masonry remodeling and new construction. Masonry additions to the original adobe pueblo included new walls building "up" from existing walls, a number of new all-masonry rooms within the original open plaza, and later new masonry rooms extending the footprint of the pueblo to the east. All of the adobe walls that we found had masonry added on top, resulting in mixed-media walls where the adobe construction extends to a variable height up from the prepared floor. In Area A, for example, we defined two adobe-walled rooms of the original pueblo,

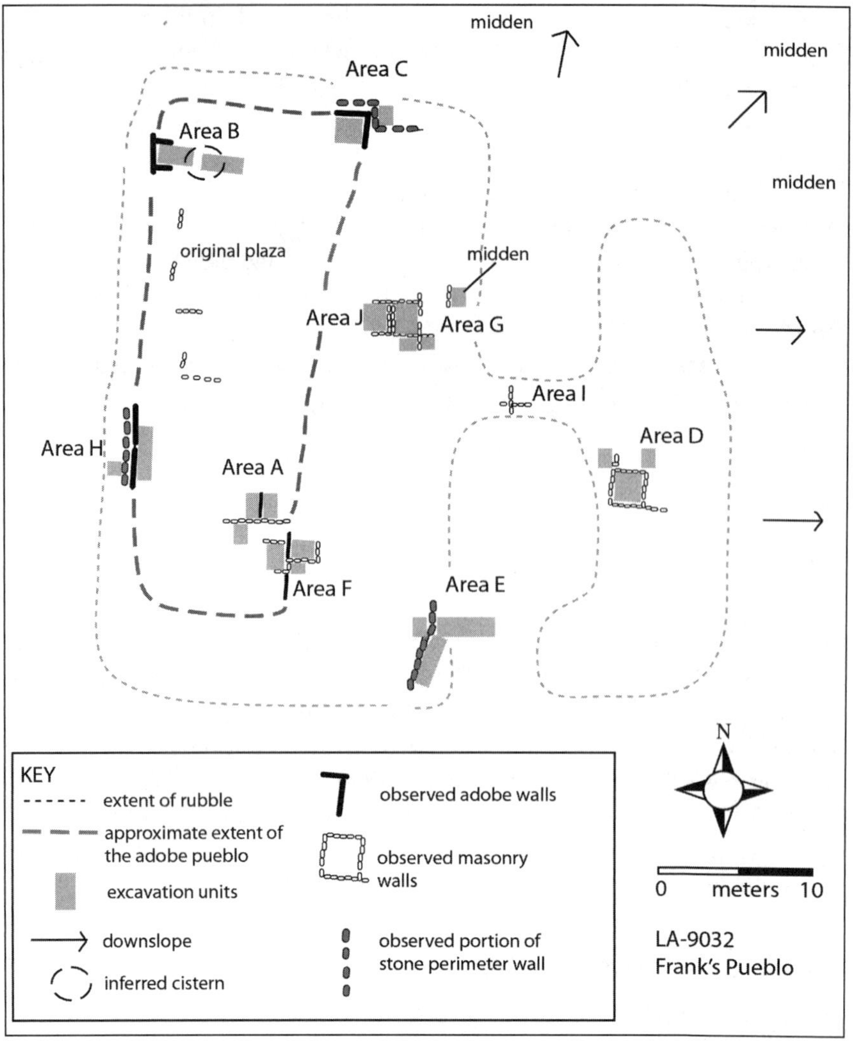

Figure 18. Map of Frank's Pueblo, LA 9032. The heavy dashed lines show the outline of the original adobe pueblo. The letters refer to excavation areas.

which had at least one doorway linking to additional rooms on the south side. At some point the original adobe wall was remodeled with masonry blocks. At the same time, masonry blocks were also used to fill in the former doorways, indicating that remodeling also involved changes in room use and interior access patterns. The floor level of these rooms remained the same, however, showing that the old adobe wall rooms were fully integrated into the new construction.

It is possible that this remodeling represents refurbishing of an aging and weathered structure; it is also possible that the masonry construction represents a rebuilding of the pueblo after a hiatus of occupation. Whatever the impetus for this upward remodeling, it is clear that the inhabitants kept the basic footprint and layout of the pueblo intact, as well as the placement of existing rooms and floors, but with some modifications of doorway access routes.

A second type of masonry remodeling resulted in new masonry construction that increased the number of rooms. Some of these new rooms were located inside the original plaza, where the walls are detectable in areas of modern disturbance (that is, exposed in pot hunter pits). Around the same time, people also added a large number of masonry rooms to the original adobe pueblo on its eastern side, nearly doubling the size of the pueblo's footprint, while retaining the basic compact rectangular shape of the pueblo. Some of these new masonry walls are abutted to the older adobe building; other rooms farther to the east show entirely masonry construction.

After these new masonry rooms were added on the east side of the pueblo, the whole construction was enclosed within a thick and well-made masonry wall. In contrast to the rather haphazard appearance of the perimeter wall at Pueblo de la Mesa, this wall at Frank's Pueblo included stones that were shaped on the exposed side and set into place with adobe mortar. Along the western perimeter of the adobe pueblo, people put the unfaced sides of very large stone blocks right up against the exterior adobe walls, leaving the pueblo exterior completely sheathed in stone. This double wall is visible on three sides of the original adobe pueblo (north, west, and south). Apparently the inhabitants decided to reinforce the original exterior adobe walls of their pueblo with masonry sheathing. Table 18 summarizes this constructional history.

The large blocky stone used for this exterior sheathing wall indicates that people invested a great deal of energy into its construction. Were they simply reinforcing the exterior walls against possible attack from outside? That is one possible idea, although an adobe wall would likely stop an arrow or spear just as readily as masonry would. Another idea is that the original adobe walls might need to be reinforced to support the additional weight of a more substantial roof, or possibly a second story. The size of the rubble mounds, however, does not seem to be large enough to represent more than one story of masonry construction.

Table 18. The history of Frank's Pueblo, LA 9032

Sequence	*Evidence/Location*	*Interpretation*
Initial construction	West Pueblo	Construction of an adobe plaza-oriented pueblo. A possible cistern was located in the plaza.
Remodeling	West Pueblo	Adobe construction was remodeled with masonry. Masonry was used to extend the walls vertically and to fill window/door openings.
Expansion to the east	West Pueblo	Partial masonry and all-masonry rooms were added to the east of the original pueblo. A thick midden accumulated to the east of Area G.
Fortification	Thick encasing wall surrounds the enlarged pueblo	On the west side, the wall was constructed parallel and adjacent to the old adobe wall; on the east side, it enclosed masonry rooms.
Catastrophic abandonment and fire	West Pueblo	Every area tested showed signs of intense burning and catastrophic abandonment.
Return to the pueblo	West Pueblo plaza and Structure 6	Large burned and unburned sherds were deposited into the cistern in the plaza. Two vessels were placed into roof fall of Room 6 before the walls fell into the room. The time elapsed between these two events is not clear.
Reoccupation	East Pueblo	Room blocks in Area D and Area I were constructed.
Abandonment of East Pueblo	East Pueblo	Area D and I were abandoned, although the sequence of abandonment is not clear. There is possible evidence of hasty departure of the inhabitants from Room 2 (Area D). Room 2 also showed some evidence of burning.

These different remodeling episodes that added living space and reinforced the exterior walls seem indeed to have been defensive strategies—unfortunately, however, they proved to be unsuccessful ones. As the field season progressed, we had yet another surprise: we realized that the entire place had been thoroughly and completely burned. It was in fact a little unnerving to imagine people apparently fleeing for their lives, leaving their household goods, their baskets of seeds, tools, and grinding stones on the roofs; their pottery water storage jars on the floors; their stores of shelled and unshelled corn—their homes—behind. At least, I hope they were able to flee; anyway, we did not find them.

Fire and Its Aftermath

In both rooms of Area A within the oldest portion of the site, a thick layer of burned roofing material lay directly on top the fire-reddened adobe floor. In one room (Structure 4) the burned remains of at least three complete large Corona Corrugated brown ware pots and a complete metate were found on the floor below the roof fall. In the other room (Structure 5) a bone scraper, a projectile point, and a mano lay on the floor below burned roof fall. Other tools such as a polished axe made of lava were found in the upper levels of burned roofing material, as if abandoned where they were used or stored on the rooftop working areas.

Excavations in the original plaza revealed a trampled surface, and also the edge of a large deep pit, which we interpreted as a cistern. The pit was roughly circular, with vertical sides, and at least two meters in diameter. Within the cistern's cultural fill was evidence for multiple, discrete fill events that resulted in a complex sequence of deposits with both burned and unburned artifacts (including some very large sherds) in the same context.

This complex arrangement of cultural fill within the pit seems to represent purposeful cleaning up that took place after at least one burning episode at the pueblo. The juxtaposition of both burned and unburned sherds suggests that people returned, or at least revisited, the pueblo after a fire; they removed charred debris from burned areas and discarded the broken pots and chunks of building material into the cistern, together with other, unburned debris from other locations.

Rooms on the north side of the site (Area C) yielded even more spectacular evidence of extensive burning. The fire was extensive and catastrophic, affecting both the interior of the room and the rooftops. Seeds were charred inside their ceramic containers, and burned maize cobs and kernels were ubiquitous. In one room, burned roof beams also attest to widespread and thorough destruction of the room and its contents. Above the beams, carbonized basketry, burned corn, and scorched and blackened ceramics record activities on the roof top that were interrupted by the fire. When the roof beams burned through, the items on the roof fell with it, landing on the floor below. Inside the room, the roof's fall smashed ceramic jars and tools where they had rested on the floor. A cache of thirteen whole projectile points, likely originally gathered into a small bag, had been tucked into the ceiling inside the room; when the roof collapsed onto the floor, these points remained protected from the fire within the roofing material.

Another burned room in Area G had been a storage room and was filled nearly knee-deep with shelled corn. Whether these two burned rooms represent one widespread fire or two different events, the people of Frank's Pueblo apparently gave up the idea of cleaning up and rebuilding these particular rooms. At this time, it appears that occupation of the pueblo shifted slightly to the east. Here, people constructed at least two linear room blocks that were located outside the heavily built exterior walls of the West Pueblo.

Building and Rebuilding in the East Pueblo

Outside of the protective walls of the original adobe and masonry pueblo, we discovered evidence of two linear room blocks (in Areas D and I), which we referred to collectively as the East Pueblo. Each linear room block was two rooms wide and about six or eight rooms long. The two room blocks were oriented at right angles to one another, creating a cleared area in front that was also bounded by the east perimeter of the West Pueblo. Both room blocks of the East Pueblo exhibit the same ladder construction that we encountered at Pueblo de la Mesa, with the long walls constructed first, and the space subdivided by secondary walls.

Test excavations in Room 1 (Area D) revealed a well-preserved floor, with adobe lapping up onto fire-reddened walls. No roofing material was preserved. Features in this room included a hearth and a stone niche with a polishing stone and mano placed within. It seems the people who used Room 1 also made a hasty departure; they left behind a number of well-preserved tools that had fallen onto the floor: four bone awls and two manos. Given this evidence, as well as the evidence of burning elsewhere in the pueblo, it is entirely possible that the room was also abruptly abandoned at the time of the fire in this area. If the roof burned and collapsed, it has long ago eroded away.

All excavated areas of Frank's Pueblo show signs of extensive burning and abrupt abandonment. It is not clear whether the two parts of the site—the West Pueblo and the East Pueblo—were occupied and abandoned simultaneously, or in sequence. It is possible that the East Pueblo was built and occupied contemporaneously with the masonry constructions in the main part of the pueblo (the fortified West Pueblo). If so, why were these room blocks not included within the encircling wall? This spatial relationship could indicate that people of the East Pueblo joined others at Frank's Pueblo later in its occupational history—that is, after people of the

West Pueblo invested the effort to build the exterior encircling wall, and apparently had no more room for accommodating more people within the West Pueblo itself.

It is also possible, however, that the constructions represent group activities at two different times. If in fact the two room blocks located outside the encircling wall were built after the major fire in the West Pueblo, then the burning recorded by the fire-reddened walls of Room 1 represents yet another later episode of burning and abandonment.

It seems unlikely that the time interval involved between these two postulated burning episodes would be long enough to be differentiated by ceramic type differences or by radiocarbon dating—after all, we may be dealing with time intervals of less than a decade or so. Right now we can only propose a temporal sequence of events happening before and after the major fire(s) recorded in the fortified West Pueblo. Some other, more sensitive means of relative dating for the site is needed to distinguish between these several possible histories of the East Pueblo.

Community Organization at Frank's Pueblo

The archaeological evidence from Frank's Pueblo ended up presenting us with more of a Pompeii scenario (Binford 1981) than we had anticipated, which has its own advantages, of course, but makes it very difficult to compare with sites such as Pueblo de la Mesa where there is abundant evidence of everyday life from deep midden deposits. However, even if the depositional contexts are not exactly comparable between the two sites, we can still make some general statements about the community at Frank's Pueblo. While the evidence must address regional concerns such as conflict and possible war, we did learn quite a bit about the internal organization of the community as well.

Evidence of Social Commitment to Place

Evidence of Jacal period structures and the buried adobe pueblo attest to the continuity and long duration of human occupation of this particular locale. Of course, we do not know if these occupations were literally continuous from season to season or year to year. For example, the observed remodeling of the adobe pueblo could have occurred as a matter of maintenance, or it is also possible that people rebuilt eroded walls after some period of absence. Similarly, the observed cleaning activities that are recorded in the cistern fill point to one or more attempts at reoccupation of the West Pueblo in spite of the apparent dangers. All of these observations represent pieces of evidence attesting to people's longstanding occupation of the site area.

This temporal continuity of occupation contrasts dramatically with the situation at Pueblo de la Mesa, where there is no evidence of Jacal period occupation. While the frontier model of settlement may thus apply for at least one of the eastern pueblos of Salinas near Jumanes Mesa (Stuart and Gauthier 1981), the pueblos in the southern portion of Salinas exhibit much greater time depth and continuity of occupation during this time of development of early villages.

Evidence of Residential Groupings

Another aspect of temporal continuity concerns the size and composition of social groups within the Salinas village sites. The basic ideas, assumptions, and social relationships that we first saw at Kite Pueblo are not unique to that one group, but were more broadly shared within this region, including the residents of the adobe pueblo at Frank's Pueblo. Again, while we do not know how many known masonry sites have a buried adobe pueblo beneath them, we have established that Kite Pueblo is not a unique case. The fact that the two adobe pueblos are similar in overall size and layout as well as construction also argues for a relatively high level of agreement and standardization of ideas regarding group size, organization, and communication within and also between residential groups across the southern Salinas region.

Evidence of Social Differentiation

One interesting aspect of Frank's Pueblo is how efficiently the inhabitants accommodated an obvious influx of population. The additional roofed residential and storage space, both within the footprint of the pueblo and also adjoining the existing structures, was obviously a rapid response to some social change. We do not know, of course, which of these constructions occurred first, or whether these two strategies for expansion occurred at the same time or in sequence. What it does tell us, however, is that the socially connected community based at Frank's Pueblo included some on-site residents and also some individuals who apparently did not reside within the original footprint of the pueblo.

This sequence of construction suggests that not all members of the pueblo village community were in fact usually in residence at the village site, the pueblo itself. We do not know if they lived nearby in more dispersed homesteads, or even possibly in some nearby jacal structures, located within a stone's throw of Frank's Pueblo. This issue highlights the difficulty in dating and interpreting jacal structures and sites. The wattle-and-daub architecture is perishable, and jacal rooms are rarely filled with anything other than windblown sand. In many cases, widely scattered surface artifacts in sheet midden provide the only available dates for occupation. Thus, while we know that Jacal period sites have jacal structures, we also know that not all jacal structures date to the same time period—for example, Pueblo de la Mesa had some jacal rooms that likely functioned as porches. It is therefore possible that some of the jacal structures visible around Frank's Pueblo may have been occupied at the same time as the main pueblo and that the occupants moved into the main pueblo when danger threatened. The exact relationship between the Glaze A pueblos and the nearby jacal sites, however, is still an open question.

We do know that masonry construction at Frank's Pueblo essentially doubled the size of the original adobe pueblo. If indeed groups of nonresidents moved into the pueblo seeking refuge, they did not anticipate leaving again. The standardized nature of the construction on the east side of the adobe pueblo seems to argue against a scenario in which individual families or small groups arrived one by one over time, each one negotiating admission to the established group. Rather, it is more consistent with the wholesale in-migration of a social group that was of similar size to the inhabitants of the adobe pueblo, such as Kite Pueblo. However, it also seems significant that this incoming population was accommodated within the village site itself. Thus, while the population may have included two social groups—the Frank's Pueblo residents and "others"—it seems that people made little distinction between the two groups when they merged into the expanded village.

Evidence of Social Integration

The most telling piece of evidence about the relationships between these two potentially separate social groupings at Frank's Pueblo is that the substantial encircling wall encompassed the whole enlarged pueblo. Whatever the exact relationship between them, the original residents coped with this population influx by incorporating the newcomers into the existing social group, and apparently simply redefined "the pueblo" both architecturally and socially. Subsequent construction of the encircling wall represents a statement that the newcomers as well as the established residents did see themselves as a single unit, at least in terms of their relationship to this outside threat.

These actions indicate that the residents and the incomers saw themselves as closely related groups, possibly as different and coequal segments of a larger social unit. This kind of evidence further suggests that pueblo sites within the Salinas region may not have housed the whole pueblo community. Instead, this evidence presents us with a scenario more similar to that of a modern small town considered together with its outlying farms and houses in the nearby countryside. In fact, one possibly similar analogy might include the historic Euro-American town of Gran Quivira, where the community included a village site cluster and also individual householders who lived farther away. The town, however, represented the village community as a whole, and housed many of the social and ceremonial events that concerned the larger social group and formed the symbolic "home" for self-reference, identification, regular social interactions, and also practical support.

Regional Relationships and Evidence of Conflict

Excavations at Frank's Pueblo record the likely presence of intergroup hostility and raiding and the strategies that people used to cope with regional conflict. Intergroup conflict is not unexpected for the region as a whole, since these pueblos along the mesa edge have long been interpreted as occupying good defensive locations, and Pueblo Seco is clearly located and constructed to maximize its defensive potential. The degree to which defensive needs affected other pueblos in the area, however, turned out to be much greater than I had anticipated.

The problem is that the archaeological concept of "defensive site" glosses over all kinds of different variables, including site location, settlement clustering, and site configuration (see Fowles et al. 2007). It is clear that Frank's Pueblo is not "just" a defensive site. Nevertheless, it is also clear that the occupants' defensive needs did play a significant role in its history. Serious discussion of this history must involve some discussion of the cultural context of conflict, considering, for example, issues involved in the anticipation and preparation for hostilities, actions intended to avoid hostilities, activities that people refrained from doing, and the highly variable nature of hostile acts in relatively small-scale societies.

Considering the evidence for defensive positioning at Frank's Pueblo makes more sense when we can break down these general ideas into specific variables and their archaeological expectations. For example, Solometo (2004, 2006) utilized ethnographic evidence of warfare and conflict in small-scale societies to identify variables associated with different dimensions of conflict (table 19; see also Keeley 1996 and R. C. Kelly 2000, 2005).

Treating each of these various dimensions of conflict in detail represents a much more detailed discussion than is appropriate here. It is clear, however, that many of these dimensions of conflict can be investigated at least to a modest extent with the limited information available from this one site. For example, we might note that the Glaze A pueblos along Chupadera Mesa and Turkey Ridge are fairly evenly spaced, with many of them seemingly aligned in line-of-sight view to one another. Beers (2012) has recently established that in fact these pueblos near Gran Quivira were in fact visually linked with one another. This kind of spatial positioning is consistent with that seen among people who are cooperating to provide an alarm system against some common enemy. Evidence of continual efforts to increase living space "inside" the walls and the labor investment entailed in constructing the heavy exterior wall indicates that this hostile, or at least watchful, environment continued for some years, even if actual acts of conflict may have been rare.

A vast ethnographic literature on small-scale warfare and violence forms the background to this archaeological discussion (e.g., Kelly 2000, 2005), as well as research on interpreting evidence of violence and warfare in the archaeological record (e.g., Arkush and Allen 2006), and specifically in the American Southwest (e.g., Lekson 2002; Nichols and Crown 2008, among many others). Here, I consider explicitly what the concept of defense might entail for small-scale groups at this time. Following Solometo (2006), I consider six dimensions involved in evaluating conflict: social distance, social scale, tactics, goals, frequency, and duration. Table 20 provides a brief explanation of these dimensions and general examples of the different variables involved. The concepts and conclusions presented here owe a great deal to ongoing discussions among all of us involved in recent research in Salinas (Julie Solometo, Matthew Chamberlin, and myself) and build on work presented in Chamberlin and Rautman (2009), Rautman and Chamberlin (2008, 2010), and Chamberlin et al. (2011).

Table 19. Dimensions of war in small-scale societies

Dimension (variable)	*Explanation*	*Factors to consider*
Social distance	This variable considers possible relationships and connections between groups.	Do the combatants share any mutual interests? To what extent do they share ideas about the conduct of war? This issue involves considering past, present and also anticipated future relationships.
Social scale	This variable considers the size of the groups that are engaged in conflict.	Do groups recruit warriors based on kinship or other obligations? How do others become involved?
Tactics	This variable concerns the details of the organization of actual conflict: the size of the specific war party, weapons, advance planning, treatment of captives, and other considerations.	How are combatants supplied and organized? What kinds of attacks do the defenders seem to be anticipating? Are non-combatants differentiated from combatants?
Goals	This variable refers to the motivations of the participants and the intended results of conflict.	Is there evidence of territorial expansion or population replacement? Are resources taken for reuse, or are resources destroyed?
Frequency of conflict and predictability of conflict	This variable concerns possible evidence of repetitive conflict, as well as the overall time frame involved in hostilities.	What would be the costs and benefits of the investment in fortifications or other labor-intensive constructions? Is there evidence of investment in cultivating allies or using spies or scouts?
Duration of conflict episodes	This variable addresses specifically the time frame of individual attacks.	Are individual episodes of conflict discrete and short in duration? Or is there evidence of extended periods of siege?

From Solometo (2004, 2006); see also Hayden 2000 and Otterbein 2004

Table 20. Dimensions of war expressed at Frank's Pueblo

Evidence found at or near Frank's Pueblo	*Dimension of conflict*	*Inferences*
Sites are clustered along the edge of Chupadera Mesa and Turkey Ridge.	Social scale of conflict	These sites may have functioned as lookout areas along social or territorial boundaries. The clusters may represent political alliances. Social tensions may have existed between the different site clusters or with a common enemy.
Pueblo Seco was one of the earliest pueblo sites. It is located on an isolated mesa and commands a good view of Arroyo Seco to the west.	Social distance between groups	Pueblo Seco may have functioned as a lookout for enemies approaching from the west. Alternatively, it could have functioned as a retreat/refuge from enemies located to the east. The anticipated enemy may have been external to the Salinas area, at least at that time.
These early pueblos are visible to one another (see Beers 2012).	Tactics of conflict	The pueblo inhabitants were concerned with communicating with one another and coordinating their actions.
The upland areas were occupied for a long time before pueblo construction, but at some point people invested in specific defensive construction.	Predictability of conflict	There was a change in people's assessments of possible conflict. Conflict (or threat of conflict) became something that was anticipated.
Defensive features at Frank's Pueblo include the encircling wall and loophole, the structured entryways, and the internal cistern (for water storage).	Duration of conflict	The need for defensive preparations were long lived. People expected to defend the pueblo itself from attack.
Continued remodeling expanded the size of the pueblo and fortified the exterior walls. There is evidence of rapid abandonment of rooftop work assemblages	Frequency and predictability of conflict	The possibility of conflict was apparently increasing. Non-residents who were affiliated with the pueblo may have moved into the pueblo itself, possibly for security. Attacks were expected to occur, but may have occurred without much warning.

Table 20. (continued)

Evidence found at or near Frank's Pueblo	*Dimension of conflict*	*Inferences*
There is evidence of widespread and repeated destruction at Frank's Pueblo. Storerooms of shelled maize were burned.	Goals of conflict	The attackers attempted to force people to leave their homes and land, not just obtain supplies. Alternatively, the pueblo inhabitants may have deliberately burned their supplies as they left, preventing the invaders from using the food and facilities.
There is evidence of rebuilding and reoccupation.	Goals of conflict; social distance between combatants	The evicted group (or at least a group with similar ideas) rebuilt the pueblo, taking into account its previous architectural organization.

My interpretation of the excavation data from Frank's Pueblo (above) uses the logic outlined in Solometo (2004, 2006). My interpretations presented here are based on conversations with Matthew Chamberlin and Julie Solometo during and after the field season. Some of these interpretations were included in slightly different form in Chamberlin and Rautman (2009).

Evaluating the Social Distance Between Groups Involved in Conflict

Social distance refers to the relationships between individuals and groups involved in conflict (Solometo 2004, 2006). In small-scale societies, this dimension includes issues of identity and group membership. In this case, close social relationships and continued interactions among the different pueblo sites around Chupadera Mesa are shown by shared use of clay sources for pottery manufacture (Chamberlin 2008:209) and also by a highly uniform style of decoration within Chupadero Black-on-White vessels across the area (Clark 2006; Chamberlin 2008).

Other social divisions are more visible at a larger geographic scale, for example, between the different clusters or divisions that Mera (1940) noted within the Salinas region. At an even larger geographic scale, we know that there were significant social and political divisions between the Salinas people and the nomadic Plains groups (Spielmann 1991). Spanish records show that the large Salinas pueblos traded maize and cotton blankets for Plains products such as bison meat and hides with Plains hunter-gatherers such as the Apache. At other times, however, the Plains groups were prepared to simply take what they needed. In fact, Apache raids are commonly cited as one factor in regional abandonment in the late sixteenth century (see discussion in Hickerson 1994).

The relevance of this historic-period relationship to ancient times, however, is quite problematic. It seems the earliest and most obviously fortified settlements in Salinas are located on its western side. Here, along the Rio Abajo, about fifty kilometers away, there is a "sevenfold increase" in population associated with evidence of a shift to upland settlements there (Marshall and Walt 1984:137). Although we do not know to what extent these rapid changes along the Rio Abajo appeared threatening to Salinas groups, there is evidence that people in Salinas made an effort to monitor and reinforce this western border, first by constructing early lookout sites such as Pueblo Seco in the southern portion and LA 503 in the northern portion of the area. Chamberlin (2008:96) also identifies fortified architectural features at the three westernmost pueblos on Chupadero Mesa, as well as evidence of extensive burning (Chamberlin, personal communication 2013). All of this evidence suggests that the people of Frank's Pueblo and the nearby pueblos perceived a common threat.

Evaluating the Social Scale of Conflict

The social scale of conflict describes the size and composition of the groups that are actually or potentially engaged in hostilities. At Frank's Pueblo, early efforts to increase roofed space inside the original pueblo by constructing rooms within the former plaza argues for an increasing size of roofed area. Later masonry rooms added to the east side of the pueblo show a second strategy to accomplish similar goals. Given that at least some of the added rooms include both residential rooms and storage rooms, it seems clear that the resident population at the pueblo increased over a relatively short period of time, suggesting an increase in the social scale of conflict, with larger groups of people impacted over time.

Evaluating Tactical Decisions

Specific tactics involved in conflict include the size and composition of groups actually engaged in combat, the weapons technology, and the specific structures and features constructed for defense. At Frank's Pueblo, the most obvious defensive construction is the thick masonry wall constructed around the entire pueblo. It is not entirely clear what advantages were afforded by stone masonry or whether these advantages were primarily those of structural integrity and strength, impenetrability, or simply ease of construction. If access to water were an issue, certainly masonry construction would involve less of this potentially scarce resource. The unusually large size of the stone used for the encircling wall, however, points to some other factor affecting raw material selection.

Evaluating the Apparent Goals of Conflict

A group's goals and future intentions also affect the conduct of warfare. In particular, conflict motivated by social issues such as revenge, honor, or status may involve different strategies than a conflict over resources where one group seeks to remove and/or replace another. In general, warfare involving future replacement of population is much more likely to involve more fatalities, while sparing resources that may prove to be of later use. On the other hand, destruction of facilities and resources such as farmland suggests an attempt to prevent a group from returning to their homes, or to prevent an "outsider" group from moving in. The amount of destruction recorded at Frank's Pueblo points to a coordinated effort to make the pueblo uninhabitable and prevent reuse of resources such as stored maize. Fire may have been used as a weapon of war by outsiders seeking to prevent the residents from returning, or even by the residents themselves, in an effort to prevent the attackers from occupying the pueblo and using those resources. In either case, however, use of fire at Frank's Pueblo represents significant damage, and reoccupation apparently entailed investment in completely new construction of the East Pueblo outside the encircling wall.

Evaluating the Frequency and Predictability of Conflict

This variable refers to the time frame involved in hostile acts, periods of violent engagement, and the estimated likelihood of repeated episodes. Here, the general concept is similar to that of cultural coping strategies involving resource use (Minnis 1985), insofar as a successful strategy involved in coping with some stressor must generally match the scale, level, and duration of the stress. Thus, for projects that involve the investment of time and labor, we expect to see a fairly close relationship between energy expenditure and anticipated use. In particular, Otterbein (2004, 2009) argues that ethnographic records show that labor investment in fortifications is strongly associated with "continual war," with attacks occurring at least once per year (see also Solometo 2006:30–31 and Schaepe 2006)

At Frank's Pueblo, the labor investment in constructing the exterior wall indicates that this proposed frequency of hostile engagements was either experienced or at least actively anticipated. The sequential increase in the size of the pueblo, and the labor investment in defensive features, also indicates that the resident population anticipated an increasing scale of conflict over time. These specific features reveal that the pueblo's inhabitants were actively and continually assessing the regional political environment and were actively positioning themselves for survival within this changing social context.

Evaluating the Duration of Conflict

The duration of conflict involves the length of time of discrete events, as well as the time span involved in general hostilities. In small-scale societies, individual battles are typically fairly brief and more typically described as raids. In the archaeological record, evidence of storage facilities can indicate the use of siege as a tactic of war, and investment in such facilities can give some idea of its expected duration. In the Salinas region, however, storage facilities for maize and water are also expected to represent the usual seasonal storage strategies common to an agricultural society in the semiarid Southwest. The observed facilities at Frank's Pueblo, or even at Pueblo de la Mesa, may not represent any particular efforts to stockpile more resources than would be expected in any pueblo of that time.

The archaeological record from Frank's Pueblo is more informative for assessing the overall duration of general hostilities in the region. Long-term occupation of upland locales (Chamberlin 2008) argues against any shift in regional settlement pattern or large-scale migration from other regions, a finding that agrees with other lines of evidence (Spielmann 1996; Rautman 2000). The construction history of Frank's Pueblo suggests that hostilities began at some time after these upland sites were well-established communities. In this case, existing social groups coped with a changing political climate using several different strategies, including population nucleation and also specific tactics of defense (Rautman and Chamberlin 2008, 2010).

Population Nucleation and the End of Early Villages in Salinas

Ultimately, the people at Frank's Pueblo were not able to avoid the effects of regional conflict. There was at least one large-scale fire in the pueblo and at least one rather determined attempt to rebuild and reoccupy the site, as seen by people's construction of the two room blocks perpendicular to each other, forming the East Pueblo (Areas D and I). People living in these room blocks, however, also experienced some problems. Perhaps there were threats to their safety or perhaps there were more attractive opportunities elsewhere. At some point, however, the disadvantages of staying outweighed the costs and risks of moving, and the last inhabitants of Frank's Pueblo left their homes.

The fire, whether purposeful or accidental, did considerable damage to the West Pueblo, and we found little evidence of reoccupation. One rather poignant discovery, however, shows people's continued engagement with the pueblo in a very tangible way. Our excavations in Area G bisected a masonry room; here, we could identify three walls and a carefully made adobe floor. This was obviously a storage room rather than a residential room because quantities of shelled maize covered the floor and mounded up against the walls. When the room caught fire, the corn burned, and the room floor and walls were reddened from the heat. The charred roof beams broke and fell into the room, spilling adobe roofing material onto the piles of corn and dragging along some masonry down from the walls.

At some later time, when the fire had burned out and the dust settled, someone came back to the pueblo (see Miller and Graves 2009 for a somewhat similar situation). Whoever came back dug a neat hole into the fallen roof debris and set two new Corona Corrugated jars into the pit. These jars were about forty centimeters high, about the expected size for a stew pot, but considerably smaller than the very large corrugated jars that were found elsewhere in the site. They had been fired, but apparently never used for cooking; they were adobe-brown in color rather than the usual sooty black. The two jars and the whole pit were then filled with clean orange sand.

When we first uncovered the jars, we had some concerns that the contents might include the remains of victims of the fire, but in fact the sand was remarkably clean of visible charcoal, vegetation, or artifacts. This kind of sand forms unconsolidated deposits along the slopes of Chupadera Mesa and is readily available from the dunes cut by small arroyos anywhere nearby. It seems significant, however, that the pit was not filled with its own back dirt or with midden sediments or cultural fill from the site itself or from anywhere around in the midden. Rather, someone obviously went to some effort to bring the sand in from off-site and differentiate this fill dirt from the surrounding burned debris. Recent analysis of the pot's contents suggests that it may have held some liquid, possibly water that included some plant phytoliths. Heat damage and deformation of the plant opal found embedded in the pores of the vessel walls suggests that the room fill was still hot and smoldering at depth when the vessel was emplaced in the rubble (Puseman and Rautman 2012). This recent finding suggests that this event occurred days or weeks rather than months after the major fire.

James Snead (2004, 2008) describes the possible emotional impacts of conflict on the ancient inhabitants of another site, Burnt Corn Pueblo, and the role of people's memories in the creation of landscapes of meaning that included the ruins of pueblos that were destroyed in violent conflict. At Frank's Pueblo, this unusual feature may record the beginning moments involved in the creation of a similar meaningful place (see also Bowser and Zedeño 2009). The visitors to Frank's Pueblo would have seen the blackened beams and the collapsed roofs and would have smelled the odor of burned piñon—as we did, too, many hundreds of years later, when we reached the levels of burned roofing. The surrounding walls would have been relatively intact at that time. Only later did the standing walls slump down, eroded by years of late-summer rains, spilling masonry blocks and chunks of adobe mortar into the room. The visitors were not placing these pots into a shapeless mound of debris; they were digging into the burned roof of a room. And when they straightened up from their work, they were not looking at an archaeological site; they were surveying the remains of their own or someone else's home.

Some scholars question whether the most important, or even the only possible, approach to the past is through emotional connection and individual imagination (e.g., Wallace 2004:195; see also Shanks and McGuire 1996; Lavender 2006; and Jones 2002 for related discussions of research practice). In this archaeological feature at Frank's Pueblo, however, we see clearly the intersection between different conceptual scales of analysis. Generalized models based on cross-cultural information such as that compiled by Keeley (1996), Otterbein (2004, 2009), and Kelly (2000, 2005) certainly help us articulate and understand the range of possible variation in abstract concepts, such as the different dimensions of warfare. However, we can see also that these dimensions of warfare are not "just" abstractions—in every case, they are played out and experienced by individuals in a particular cultural context at a particular place in time and space, and they affect individual lives.

Whatever the specific circumstances were that prompted evacuation and finally abandonment at Frank's Pueblo, these events affected people in many different early pueblos across the entire Salinas region. The fire at Frank's Pueblo would have been visible for miles, and certainly the inhabitants of the other nearby Glaze A pueblos would have been aware of these events and actively engaged in monitoring and managing the aftermath. We do not know where the residents from Frank's Pueblo went, of course, but it is around this time that nearby pueblos, such as Pueblo Pardo and Gran Quivira, begin to increase in size, adding multiple rooms and room suites to existing structures, building additional room blocks, defining multiple plazas, and creating sprawling multistoried towns that were home to many hundreds of people.

By the end of the 1300s, most of the small Glaze A pueblos across the entire region were empty. The remaining pueblos apparently were able to accommodate refugee groups and develop organizational strategies that allowed them to sustain themselves and their position in the Salinas region for centuries, despite the possibility of continued threats—possibly from the Rio Abajo to the west, or the Plains to the east. By the time of Spanish conquest, the foreign administrators and mission priests described and treated these pueblo towns as distinct and fairly coherent political units, while the Indians' continued use of multiple ceremonial kivas records the continued social importance of smaller (intrapueblo) social units in at least some contexts (Hayes et al. 1981:9).

In some places in the Southwest, physical anthropologists find differences in health, disability, disease burden, and overall quality of life for different groups within a single pueblo (e.g., Stodder 2012). In the Salinas pueblos, however, as far as I know, there is no evidence of intrasite pueblo differentiation in life experiences that could be detected in skeletal remains, at least until Spanish conquest (see Hayes et al. 1981 and the chapters therein). We know that one late-arriving group of immigrants did come to settle at Gran Quivira amid a flurry of building activity from about A.D. 1545 to 1607. It seems likely that they arrived with new ideas and new burial customs of cremation. Whatever the nature of this possible in-migration, the biological and also archaeological evidence shows no particular evidence of spatial or social segregation, or differential life experiences (see Hayes et al. 1981). This conclusion does not imply that the people of Gran Quivira or any other Salinas pueblos did not recognize internal social, political, or economic divisions; the evidence that is available, however, shows that such divisions did not result in differences in individual life histories to such an extent as would be manifest in skeletal remains.

What different processes were involved throughout this period of population nucleation and in-migration? How were the Pueblo IV period pueblos able to cope with relatively large groups of arrivals over a relatively short period of time? Did the smaller refugee groups maintain their own identities within the larger pueblo towns, or did the importance of their differing origins play only a small role in this new social context? What new organizational strategies did the late pueblos develop to cope with a change in the scale of social interactions and to incorporate newcomers? Recent research in Salinas (e.g., Capone 1995, 1997; Graves 2002) has addressed some of these issues of regional interactions during the Spanish mission period. But for my research, this period of population nucleation marks the end of the time that we can reasonably classify as the period of early village communities in Salinas.

CHAPTER SEVEN

Constructing Community in Early Salinas Villages

I see the principal steps in the archaeological process—namely, investigation, analysis, interpretation, and peer evaluation—not as a closed circle but as an open spiral which, upon repetition, reaches ever forward. Without fresh data and refined techniques, we operate within a closed circle.
—Emily W. Haury, *Emily W. Haury's Prehistory of the American Southwest*

When we as archaeologists (or even as tourists) view large nucleated pueblo sites such as Gran Quivira, with hundreds of rooms and many different plazas, we do not usually ask: "Does this site represent a community?" We simply assume the people who lived there were a community of some sort. The Zuni conception of their pueblo as the "middle ant hill of the world" (Cushing 1901:7) conveys a concept of momentary population density, true, but even the most basic understanding of ants involves recognition that individuals are somehow related to the whole. Interpreting an apparent archaeological village site as a community thus necessarily involves a discussion of internal social and political relationships among individuals and groups that exist within the village community, and the roles of these individual and group relationships in affecting regional trade or political alliances (e.g., Isbell 2000). This discussion asks what aspects of individual behavior and community organization create the most highly visible and patterned activities at a given locale or site, what processes of village life were important or not particularly relevant over the long run, and how different aspects of community were expressed or minimized through each individual's participation in daily life in and around the village site itself.

These basic questions regarding the relationship between an archaeological site and the living community are perhaps more obviously relevant when archaeologists consider sites that are somehow different from most other archaeological or ethnographically observed examples. For instance, while archaeologists such as myself might be able to get away with innocently naming LA 38448 (Kite Pithouse Village) a "village," thus implying the existence of at least some sort of social organization resembling a village community, we may encounter more debate in certain cases where the intellectual or political stakes are higher. In particular, sites that seem unusual or different because of their early date or large size, such as Shabik'eschee at Chaco Canyon, rightly come under greater interpretive scrutiny (e.g., Wills and Windes 1989; Wills 1991; Wills et al. 2012).

Various interpretations of the different pueblos and apparent clusters of small pueblos at Chaco Canyon also illustrate the problem of defining and understanding the relationship between archaeological concepts of site, village site (such as "a pueblo"), and site clusters on the one hand and sociocultural concepts of community on the other. For example, Gilpin (2003) reviews how different archaeologists have used various concepts of "community" in reference to the archaeological sites of Chaco Canyon and also more generally. At Chaco Canyon, archaeologists also debate the extent to which these ideas are appropriate for thinking about the unusually large Great Houses. Lekson (2009), for example, argues that these large pueblos may not fit well into a model of what I would call the village community, but are better envisioned as special purpose buildings such as palaces or regional ceremonial centers. A similar argument concerns the interpretation of the smaller pueblos in Chaco Canyon; again, because of the unusual contrast in size of the sites with nearby Great Houses, archaeologists make more of an effort to consider explicitly issues of the timing and continuity of the occupation of the small-unit pueblos, the identity and social status of the residents, and their relationships to people in the Great Houses and within the canyon as a whole (e.g., Kantner 1996; Plog 1997). A similar set of archaeological discussions focuses on interpretations of Casas Grandes, another very large pueblo in northern Mexico (e.g., Lekson 1999; Whalen and Minnis 2001).

These unusual cases remind us of the pitfalls involved in making too facile assumptions and in being too quick to draw parallels between the various ethnographically observed Pueblo Indian groups and the organization, assumptions, and way of life of their ancestors many hundreds of years in the past. Ethnographic descriptions of life within particular pueblos that were recorded by visiting anthropologists in the twentieth century (e.g., Stevenson 1904; Lange 1959) may not accurately convey the range of variation in social relationships that are recorded by the material remains of village sites in the more distant past.

The Salinas area represents a somewhat different situation in that we do have an extensive written record from the Spanish mission period that dates much earlier, to the sixteenth and seventeenth centuries. Because of people's interest in the Spanish mission sites and the history of their development for tourism in the US National Park system, we also have much archaeological evidence regarding the social, political, and economic relationships within the Spanish period pueblos, their relationships with other pueblos in the Salinas region and other areas, and also with Plains hunter-gatherers. It is tempting to assume considerable continuity of these documented relationships back into the past.

Indeed, because the Salinas region was apparently *not* experiencing extensive population immigration and because the region as a whole exhibits considerable continuity in overall population size, it may be true that the Salinas pueblos represent an unusually conservative or stable way of life. Simply because of their size, however, I would argue that these very large late pueblos represent a very different kind of village community from those we expect to see during the Early Pueblo period, almost five hundred years earlier. We cannot necessarily assume that the social processes and relationships that characterize these very large pueblos had a time depth of this magnitude. In this chapter, I review the extent to which our basic understanding of early village communities (as per Bandy and Fox 2010) helps us understand early village societies in Salinas, and the strategies that people used to cope with changes over time.

Interpreting Early Village Sites in Salinas

In the Salinas region, early pithouse village sites occur relatively late in time compared to some parts of the Southwest. In this particular area, therefore, concerns about the domestication and adoption of maize agriculture and the association of these subsistence changes with village origins and sedentism are just not relevant (see Wills 1988a, 1991). In fact, the basic economic pattern of farming along with hunting and gathering wild foods remains fairly constant across the entire archaeological record considered here. This study of early village and community can therefore focus on other concerns, such as the social and political organization that we see particularly manifest by the features of the built environment. As it happens, my archaeological fieldwork has emphasized excavation; I simply assumed that archaeological sites with structures represented village communities, and I focused on other issues. Only later did I begin to question this linkage more explicitly. That is, when we see what seems to be an archaeological village site, how can we tell if the people who lived there formed a community? And if so, what social relationships contributed to creating and maintaining that community over time?

Answering this question has involved discussion of definitions of community and the relationship between anthropological concepts and the archaeological record. In this volume I focused particularly on how archaeologists might use evidence from the built environment to make these conceptual linkages between material remains and the social dynamics involved in the organization and operation of early village life. I particularly considered issues of social integration and differentiation within the village community, and the ways in which architectural interpretations of rooms, room suites, and room blocks relate to dynamic social relationships that anthropologists might gloss as households, kin groups such as lineages or clans, and other forms of social segmentation (e.g., Lowell 1996; Feinman 2000; Kahdahl et al. 2004). While some researchers in the past have attempted to correlate explicitly these kinds of architectural features in archaeological sites with specific social groups identified among the modern Pueblo Indians (e.g., Hill 1982), my study is not nearly detailed enough to attempt such interpretations about the varied uses or life history of individual rooms or the social relationships of the occupants within rooms, room suites, or room blocks. The constructional history of each village site as a whole, however, does provide some indication of those social segments (groups) that were important enough to be manifest in, and reinforced by, the built environment. In the following sections, I summarize some of the issues that I considered in interpreting the archaeological evidence of early village sites in the Salinas region and that I used to reach a better understanding of the ways in which relationships of social differentiation and integration structured peoples' lives within these village communities.

Constructing Community During the Pithouse Period

Most Paleoindian and Archaic sites in the Salinas region occur as small lithic scatters (Tainter 1979) and likely represent fairly short-term occupations by small groups of mobile foragers. In Salinas, the first likely village sites—those exhibiting substantial labor investment in structures and storage facilities—date to the Pithouse period. In the Salinas area, the Pithouse period includes an early occupation dating to about A.D. 600 and a later occupation that dates after A.D. 900.

The pithouse dwellers of Salinas interacted extensively with a variety of peoples to the south, in the highland areas of the Sierra Blancas, and also to the west, around the Rio Abajo (Rautman 1993b). The general sense is that these pithouse sites represent habitual use areas, with defined task areas, living spaces, and storage spaces. Even if people did not occupy these sites year-round, they likely returned seasonally. To the extent that people reoccupied the houses, used items and food from the extramural storage features, and presumably replanted their maize fields, we can envision these social groups as being fairly temporally stable and bounded in membership. In societies with bounded social groups, groups have a continuity that transcends any given individual's participation or lifespan, and individuals have a socially recognized place in one or more groups, although their roles are expected to change over time (Rogoff 2003).

Evidence from the built environment at Kite Pithouse Village confirms that people occupied or reoccupied quite specific locales across a considerable time span of several generations. The houses at this site were smaller than those used in the earlier portion of the Pithouse period, possibly signaling a change in the seasonality of occupation or the size of the coresident group. Nevertheless, Kite Pithouse Village as a whole exhibited a high degree of internal spatial differentiation and organization, where living space is clearly differentiated from midden. Even within a single site, therefore, the activities of multiple residential groups were clearly spatially separated and segregated, and this practice was sustained over many years, if not generations.

While much of the architecture at these pithouse sites now lies below the ground surface, when people lived in them, daily traffic would have been channeled around roofed houses, empty storage pits, older unroofed pithouses, and any number of temporary and ephemeral structures such as sunshades or ramadas. We simply do not yet have the larger-scale spatial coverage at any Salinas pithouse site to determine whether village sites of this date show any defined spatial organization of structures or if there are any signs of formally defined communal activities—such as the large activity area that Chamberlin (2008) observed at some jacal sites. Obviously, there is still plenty that we do not know about community organization during the two different periods of pithouse use. Based on this reexamination of the evidence, however, at least we now know how and where to look, and how we might interpret what we find.

Constructing Community in Jacal Period Sites

In the Salinas region, use of jacal architecture for sunshades or ramadas occurs throughout the chronological sequence, but the Jacal period represents the time when jacals, rather than adobe or stone pueblos, dominate the built environment. Surface ceramic collections date jacal sites to about A.D. 900 to 1150, suggesting that some jacal sites as a whole may also overlap in time with the time of occupation of some of the latest pithouse sites.

We lack excavation data from any site of the Jacal period in Salinas, but in contrast to the pithouse sites, we have a much better idea of overall site layout. Comparable jacal villages in the southern Jornada Mogollon area at this time show much more structured use of space, with individual room blocks grouped neatly around plazas (Lehmer 1948:86, and also Caperton 1981:4). In contrast, in the Rio Abajo area to the west, sites of comparable time exhibit a less standardized site plan. These observations suggest that people in these different regions maintained a certain degree of internal continuity over time, resulting in diverging local histories (see Cameron and Duff 2008). In Salinas, the organization of jacal rooms into contiguous linear room blocks, and also clusters of room blocks, shows a significant degree of concern with spatial organization within each village site. Although there are differences between sites, within a given jacal site there seems to be little internal differentiation in room sizes, no obvious evidence of differential room access, and no "internal" rooms that would be accessible only from other rooms in a grouping. All of these lines of evidence point to a small number of organizational principles that guided construction and use of structures during the Jacal period. However, variation in the size of the room groupings within each site, that is, the length of the contiguous room blocks, indicates a degree of variation in the social definition and continuity of the room block–dwelling social unit.

The distribution of jacal sites in Salinas at this time shows a fairly stable social landscape, with jacal settlements dispersed across the region. Chamberlin (2008:231) found little overt territorial signaling in the placement of shrines, markers, or other facilities across his study area near Gran Quivira. This lack of assertive boundary marking suggests that people's control of land and resources was relatively uncontested among the different groups throughout the region.

Another line of evidence for this generally stable social landscape is the in situ development of many architectural features such as the contiguous room block and the plaza (Chamberlin 2008:231). For example, in the early part of the Jacal period, peoples' conceptions of shared communal space are variable. At one site, communal space seems to consist of a leveled activity area on the edge of the cluster of residential structures; at other sites, communal space is defined between loosely clustered linear room blocks. Later in time, communal spaces are more clearly defined and larger in size (Chamberlin 2008:234).

A continued trend toward tighter nucleation (spatial clustering) of different room blocks within a given locale, however, would potentially operate at cross-purposes with a simultaneous social trend toward large and visible public spaces. In this context, the central plaza such as that described at Kite Pueblo (LA 199) and also at Frank's Pueblo (LA 9032) represents the melding or intersection of those two general social strategies for managing space within a village. It is clear that the central plaza on early Salinas pueblo sites represents a local compromise solution to these two different trends, rather than any "new" idea about public space, or one imposed from outside groups.

Constructing Community in Early Pueblo Sites

In some respects, the early plaza-oriented pueblos made of adobe or masonry seem completely different from the scattered jacal structures of earlier village sites. However, it would be a mistake to associate this change with any dramatic change of personnel or radically new ideas. Chamberlin's (2008) survey documented this in situ development of the idea of open space during the Jacal period, and excavation at Kite Pueblo also showed that the pueblo's inhabitants were most probably "locals," already living in the area and well acquainted with one another at the time of site construction (Rautman 2000).

Evidence from the built environment does record, however, a change toward standardization of the built environment at several different conceptual scales (Rautman 2013a). Within a given jacal site, we see standardization of room size, but a range of variation in the length of the contiguous room block. In contrast, the plaza-oriented pueblo creates a more standardized linear multiroom room block. This standardization was accomplished in part by simply eliminating the concept of the small two-room room blocks, thus redefining the room block as a larger architectural and social unit. The plaza-oriented pueblo also standardizes room blocks in the sense of their location vis-à-vis the visible open space of the plaza. In the plaza-oriented pueblo, all room blocks surround and thereby define the plaza. As a further result of this spatial arrangement, moreover, every single room block is situated in the same position relative to the plaza.

In comparison to a pueblo arrangement in which room blocks of standard size might be organized in parallel lines, or scattered randomly across the village landscape (see Reed 1956), the plaza-oriented pueblo emphasizes both visually and also physically the principle of equal participation in the use of communal space. In the plaza-oriented pueblo, all room block groupings share the same kind of access to the plaza; no group is closer or farther away. In addition, each room block also shares the same position vis-à-vis the pueblo itself: each room block has one side facing the plaza and one facing the village exterior. This arrangement blurs the distinction between domestic and public activities since the rooftops as well as the plaza are thus visible to all, and even the most mundane tasks of daily life are recast as public performance (see also Graves and Van Keuren 2011).

The shift to plaza-oriented pueblo thus clearly expressed in architectural terms an emphasis on the village site as a whole, while emphatically minimizing the differences between its component parts. In some respects, this built environment also seems to overtly exclude others from the village site, clearly defining and distinguishing pueblo outsiders from insiders. For this reason, many archaeologists assume that the plaza-oriented pueblo represents a strategy of defense or protection from outsiders. However, I would argue that the plaza-oriented pueblo also has a more pervasive role in constructing community, creating a sense of "insider" group identity, and reinforcing this identity through the most mundane actions of everyday life.

Evidence from Frank's Pueblo indicates, however, that this identity was not limited only to the people who lived inside the village site itself. In fact, the plaza-oriented pueblo here seems to have functioned rather as the medieval European castle—at least in the limited sense of providing a place of refuge in a crisis. In this case, the latecomers to the pueblo may have been village community members who happened to live in dispersed outlying farmsteads, or possibly they came en masse as refugees from another pueblo. We cannot tell from the evidence available nor do we know for certain whether occupation was continuous from the time of the adobe pueblo to the time of masonry remodeling. It is significant, however, that the masonry construction at Frank's Pueblo basically doubled the area footprint of the village site. More importantly, however, the "new" construction did not relegate incomers to separate areas, but rather accommodated this population influx by adding rooms inside the plaza itself and also in attached rooms that were then encompassed within a new encircling wall. These actions show a persistent social emphasis on whole-group unity and similarity, expressed in and reinforced by the construction of the built environment.

We do have some evidence of changing social ties between village communities across the jacal-to-pueblo transition period, expressed in changing use of different ceramic clay sources. We know that the iconography of Chupadero Black-on-White is highly standardized across a large geographic area, including both the Salinas area and also the Sierra Blanca region to the south (Clark 2006; Chamberlin 2008). However, compositional analysis of pottery from these two different areas shows that production and distribution networks were fairly localized within each region, with little exchange of Chupadero Black-and-White pottery between regions (Clark 2006).

Within the southern Salinas villages, however, Chamberlin (2008) found some interesting temporal changes in peoples' use of different local clay sources in and around Chupadera Mesa. During the Jacal period, the villages in his study area used a wide variety of clay sources. This evidence suggests a fairly unrestricted and open system of access to clays across the whole southern Salinas region. Distance played some role in structuring this interaction; that is, the assortment of clay sources in pottery found on jacal sites located on the eastern portion of Chupadera Mesa was similar to that used by people at nearby Gran Quivira (located just off the eastern rim of the mesa). In contrast, jacal villages located on the western side of Chupadera Mesa showed use of a mix of clay sources similar to their own neighbors around Chupadero Arroyo, which drains west from Chupadera Mesa.

In contrast, during the Early Pueblo period, the mesa-top pueblo villagers seem to close ranks, so to speak: there is a decrease in the range of variation in the ceramic sources used. As a result, the mesa-top villages appear more similar to one another (Chamberlin 2008:156–157). At the same time, the mesa-top villages' pottery sources become clearly more differentiated from those seen off-mesa at Gran Quivira, only an hour's walk away.

It is interesting that this shift occurs at about the same time that we see a change in architecture at Gran Quivira—the construction of the circular pueblo (Hayes 1981). Because so little is known about the Early Pueblo period, we really do not know if this correlation is in fact significant and what it might mean. However, we do know that circular pueblos are rare in the Southwest (see chapter 6). In terms of the built environment, however, a circular pueblo is similar to a plaza-oriented pueblo in minimizing the importance of village site sub-units and emphasizing the unity of the village site as a whole. At the same time, plaza-oriented pueblo forms—square, rectangular, or circular—convey a sense of the plaza as "theater in the round," both providing and also enforcing a public space within which even the most trivial and mundane domestic activities are both visible and communal (Graves and Van Keuren 2011; Chamberlin 2012). Activities that take place within the plaza are also demonstrably elevated in importance, occurring in the same space that is used for ritual and other ceremonial gatherings.

The research reported here, however, indicates that the overall change to plaza-oriented pueblos in Salinas is not a break with tradition, as it seems to have been in the Rio Abajo region, but rather represents a widespread and rapid adoption of shared ideas that were solidly based in local history. If in fact the idea of the plaza and conceptions of ritual space that it represents was an idea that originated among the Western Pueblo, it was an idea that fit easily into what was already a basic pool of shared knowledge among the various Salinas village communities. This background may also help explain why, once people did begin constructing early masonry pueblos, they seemed to act virtually simultaneously across the Salinas region.

Over time, the plaza-oriented pueblo presented some disadvantages. It is hard to add onto, for one, and sometimes the interior space of the plaza had to be sacrificed to accommodate more people within the fortifiable area. It was also hard to integrate more residents into the basic structure while maintaining the same layout. And eventually, as times changed, people did look for other solutions to these problems. The multiplaza, multi–room block pueblo layout that tourists see today at the large pueblos of Salinas Pueblo Missions National Park may record one such solution to the problem of constructing community within this context of continued population aggregation.

About a century and a half later, around A.D. 1550, the archaeological record at Gran Quivira shows some relatively rapid influx of customs and ideas from the Western Pueblo traditions. Hayes (1981:12) specifically notes the appearance of some cremation burials that date to about the same time, and attributes this change to an apparent arrival of migrants at this time. However, he argues that the continuity of utility wares and the persistence of inhumation burials across this whole time span, from the earliest Glaze A occupation through the Protohistoric period just before Spanish conquest, indicates that this migrant population settled within existing local groups, rather than overwhelming or displacing them.

Later occupants at Gran Quivira also changed the way that they organized their built environment. Instead of building standardized rectangular room blocks, they accommodated population growth by adding rooms and groups of rooms on to the existing architecture, creating what may have been interconnected "room suites" within sprawling "houses" that included over two hundred contiguous rooms (Hayes 1981:figure 16, p. 26). Other room blocks of contiguous rooms have not been excavated, but can be identified as large mounds that are clustered around various open plazas. A similar sprawling organization of the late sites, with many mounds of varied size and shapes and many plazas, also characterizes the other large pueblos of the area. At Gran Quivira and the other large pueblos, larger and smaller round kivas indicate a different kind of architectural expression of ritual and social space that occurs much later than the time of early village communities that I consider here.

Constructing Community in Early Village Society

In this volume, I use the archaeologically visible evidence of the built environment to help understand the development and organization of social communities. Even the most mobile foragers create some sort of built environment within an overnight campsite or resting spot where the organization of structures, features, and activities areas is hardly random (Wiessner 2002). Wherever people remain for any period of time, they organize their activities in some way, and this use of space does in fact affect social life and expresses certain assumptions about social relationships (Kent 1990, 1991).

In an area occupied for longer periods of time, the location of structures and features affect walking paths and lines of sight and the definition of inside and outside or may designate spaces for different activities. These patterns create a new cultural landscape of the built environment. Very mobile foragers must deal with the significant challenge of maintaining social ties over time in the absence of face-to-face interaction; as people are more sedentary, the issue of "too much togetherness" becomes increasingly important. Structuring people's daily life by means of the built environment is one way of managing the social stress of daily interactions (Johnson 1982, 1989; Gaines and Gaines 2000; Bandy 2004; Stanley and Willits 2004).

The archaeological remains at a given village site represent a sampling of these decisions. As occupation duration increases, people commonly make greater investment in durable construction of some facilities (Binford 1982) and segregate some tasks from others. Village sites represent the built environment within a larger local cultural landscape, which may include middens, dumping areas for wastewater and human waste, small garden areas, animal pens, or seedbeds (Doolittle 2004; Vlasich 2005). Within a village itself, the built environment can express and also reify power differences, symbolic references, and cosmological schemes. The expression of these same differences can also be minimized, downplayed, or suppressed in the organization of buildings and social spaces.

Building materials may derive from the existing environment, but buildings also create new microenvironments of sun, shade, and wind; disturbed soils and middens attract various plants and animals, and human actions affect drainage patterns, ground cover, and other local, small-scale geomorphic changes. At a larger spatial scale, of course, the cultural landscape includes the built environment of field houses, shrines, agricultural features, and other markers, as well as trails, specific resource areas, extraction sites, and other meaningful places (e.g., Snead 2008).

Whalen's (1981a) study of early Southwestern village sites across the pithouse to pueblo transition notes a general correlation between regional population size increase—an increase in social scale—with a change in the organization of society, or structural scale. In fact, there is considerable evidence that development of a relatively formal organization for exchanging information and making decisions at a social scale larger than the household does represent a density dependent phenomenon. We do see that as the size of the interacting group increases, these sorts of tasks that were accomplished before by individual and informal negotiations tend to become more formal—that is, tasks become increasingly differentiated, standardized, and explicit (Johnson 1982).

In modern practice, we all have had experience with an increase in bureaucratic formality: a group's meeting times may become more regularized or scheduled, and written bylaws may specify group membership, organizational structure, and the process and scope of decision-making. In all these respects, there is a trend toward more standardization, more organizational structure, and more specifications; fewer parameters are left undefined, and fewer decisions are left up to the individual. In discussions of village society, these ideas of differentiation and integration include groups at two different scales: within the village settlement as a whole and within existing village subgroups such as households and/or corporate kin groups.

Issues of Internal Differentiation

How small communities cope with internal differentiation while at the same time maintaining a sense of identity as a whole lies at the basis of an ongoing discussion of the role of corporate groups among early village societies in the Southwest and the relationship of this process to architectural change (Brew 1946; Wheat 1955; Bernardini 1998; Altschul and Huber 2000; Peregrine 2001; Schachner 2010). Feinman (2000) and Feinman et al. (2000) specifically relate architectural change in early villages during the pithouse to pueblo transition as the physical expression of a shift to corporate decision-making organization. In these models, not unique to the Southwest, archaeologists consider the contiguous room block as the material expression of a cooperative group that is larger than one nuclear family (Whalen 1981b; Byrd 1994, among many others).

Schachner (2010) has recently explored how the development of corporate groups, composed of several cooperating families, might be expressed in the architecture of early Southwestern villages. He argues that corporate group formation would be expressed in, for example, increased residence size and storage capacity and increased formality of activities, including differentiation of space to separate storage, living, and specialized household manufacturing areas.

As Hammel (2005) noted, corporate group formation represents primarily a mechanism for nuclear families to cooperate in organizing labor. In the American Southwest, we tend to assume that these corporate groups are kin-based, extended family households (Lightfoot 1994), possibly an early form of the clans that are described among historic and modern Pueblo Indian groups (Peregrine 2001; Ware 2002). Many Southwestern archaeologists seem to assume that this differentiation existed in a competitive environment that emphasized corporate group unity at the expense of village-wide solidarity. Whether this proposed differentiation and competition was primarily economic (Schachner 2010) or ritual (e.g., Lekson 2006a, 2006b), this emphasis on intravillage social differentiation is associated with a change in the size of the cooperative, food-sharing unit. According to this general model of early Pueblo Indian society, corporate groups consisting of several related nuclear family households formed the unit of food production, storage, and sharing.

According to this logic, I suggest that the contiguous room block can also serve as an integrative mechanism for each resident group. Although each member family may differ in size and composition across the life cycle or through random events, the standardized room and contiguous room block reminds every resident every minute in architectural form that although individual group members may operate relatively independently in some contexts, their lives and livelihoods are deeply interconnected and interdependent.

Issues of Village Integration

At the scale of the village community itself, however, these corporate groups can also be expected to differ from one another in social variables such as size, demographics, knowledge, and other attributes of social capital. These relationships are also temporally variable, as different groups may become larger or smaller, or more or less politically or economically successful over time. And to the extent that corporate groups were economically and socially independent, if not competitive, their formation as village factions could pose an additional challenge to village unity as a whole.

Archaeologically, village integration is often manifested in the material record with specific standardized symbols or colors, religious or secular group rituals, ceremonial activities such as feasts, standardized craft manufacture (Bartlett and McAnany 2000), and other means of expressing, enacting, and enforcing group membership. Integrative mechanisms thus include ideas and activities that create, express, and invest group identity to its members. Integration of groups at this social scale may also include incentives to encourage people's cooperation; on the other hand, less attractive forms of obligatory integration can involve penalties for resistance. Where integration is institutionalized, activities may take place in specified group areas, especially if the groups involved become large enough so that they cannot be hosted in any one individual's household space. In some cases, the social benefits of a large group experience may even justify the added labor, expense, and planning involved in building specialized facilities.

In the northern Southwest, Ferguson (1989) points out that the basic early Prudden Unit Pueblo (shown in figure 4, p. 61) reinforces social and political interconnections at the scale of the village site itself. The repetition of residential units provides an everyday framework of interdependence and unity among the individual parts. In this respect, the basic Prudden Unit Pueblo represents a visual depiction, and daily experience, of mechanical solidarity (sensu Durkheim 1933)—a structural repetition and reiteration of functionally equivalent social elements (rooms). Using similar logic, one can argue that the whole aboveground room block thus represents an integrative structure, in and of itself, regardless of the presence of any special communal space or construction.

At the conceptual scale of the village, therefore, development of jacal villages in Salinas may represent the beginnings of a change in structural scale affecting the organization of a resident group, and the relationships between resident groups and the village as a whole. We do not know if there is any spatial clustering of pit structure within a settlement, but we do know that intrasite spatial clustering is a dominant organizing principle for the jacal villages. If the room block does represent a cooperating social group, in which people in different room blocks may have quite different resources, social capital, and life experiences, then the idea and challenges of social integration move to a different social scale—that of the village as a whole. Just as corporate kin groups are not simply very large nuclear families, so village organization represents a different set of social relationships than those that operate within a single large lineage or clan. Principles of village integration thus involve fostering group identity, loyalty, and cooperation at a larger social and also structural scale.

In any society, we can expect people to identify with multiple and varied groups, including the nuclear family, the extended family or lineage, age grades, task groups, interest groups, and so on. Many of these groups will be local in spatial scale, while others may be quite spatially extensive. In modern times, different nation-states face similar problems in persuading people to identify first with their country, and secondarily with their local village, tribe, or traditional ethnic group.

If we assume that social communities are defined partly by self-identification and some sort of internal community cooperative action, we should look for signs of village-level integration. But when does village-level integration and unity become *the* most important concern for our archaeologically defined Salinas villages, overriding the expression of different subvillage social groupings? When do we see signs that people within an archaeological village site show significant concern with social integration at the scale of the village as a whole?

If we think back to the first jacal structures in the Salinas area (see chapter 3), Chamberlin already found evidence of "group activity areas" and site planning during the Jacal period (Chamberlin 2008:96). I had argued earlier (Rautman 2000) that the adobe pueblo builders would have to have been already acquainted with one another and sharing ideas about domestic space at the time when they began building these first plaza-oriented pueblos. Now we know that some of those inferred social connections are already present in Jacal period sites.

But is this kind of village-level organization possible only for groups using contiguous surface rooms that can be made into room blocks? Obviously, people's actions and the principles of group organization are not necessarily connected with any one kind of architectural form. However, given the observed importance of room block organization in later Salinas village sites, we might ask whether this kind of architectural expression is unique to sites with surface rooms. Do we know of any pithouse sites that show any evidence of group space, integrative spaces, or group activity areas? In fact, yes. In western Texas, Scarborough (1989) found evidence for a centralized activity zone within the overall spatial organization of an early pithouse site. He argues that this reserved clear activity area shows interhousehold cooperation and coordination of activities, that is, some sort of corporate organization at the social scale of the village, even at this very early period in his area. We therefore know that this particular kind of social organization need not coincide with any particular type of residential structure, architectural form, or time period.

In the Salinas area, however, we simply do not have the spatial information required to make this assessment for pithouse sites. Site organization is hard to evaluate in the Pithouse period because the house pits are filled with windblown sand and colluvium deposited at the base of the mesa. Unlike some other areas in the Southwest, the pit structures are therefore not apparent on the ground surface. Whalen (1990) obtained good results using a soil auger to define structures; I used a similar technique to find structures at Kite Pithouse Village, but it was quite labor intensive, and the results were ambiguous. Heyman (2004) also attempted to do some subsurface mapping at LA 38448 using ground-penetrating radar, but with limited success. Stokes and Roth (1999) have had better luck in using ground-penetrating radar in their area, so we might consider trying again.

At any rate, we do know that our Salinas people in the jacal sites were already cooperating and organizing themselves to accommodate one another in their use of space. Thus, to some extent, people living in the different room blocks within jacal sites did view themselves as connected with people in other room blocks as part of some kind of social community.

In the Salinas region, however, a second kind of organizational change is expressed in architectural form later in the archaeological record. Although there is no evidence of external in-migration or even significant local population movement, people throughout the Salinas rapidly reorganized their existing room block units into plaza-oriented pueblos during the Early Pueblo period (see chapter 4). This organizational form clearly defines "the village and its residents" and separates each village from its neighbors; it may have had some defensive advantages as well. The plaza-oriented pueblo is not unusual in the Southwest, but neither is it a ubiquitous architectural form.

Early villages in Salinas thus record the effects of two conceptual scales of integrative repetition: repetition of room size as well as repetition of rooms into contiguous room blocks. In the Jacal period, the overall linear form of the room block shows a degree of group consensus in the organization of social groupings, but the variable length of these room blocks show a certain amount of social latitude in the conception of "the room block–using group." At a larger scale, repetition of these differently sized room blocks across the locality created a single settlement, a cluster of room blocks. Across the Salinas area, these jacal settlements show a broadly shared regional concept of settlement organization among local groups who already had a history of occupation in specific locales. Local variations in site layout indicate that within these general shared parameters, local groups modified their built environment as needed to accommodate their specific circumstances.

The transition to a plaza-oriented pueblo in the Early Pueblo period standardized and simplified this local variation. The plaza-oriented pueblo involved a change in the structure of the room blocks themselves, creating room blocks of standard size. To the extent that room block size and overall shape are related to social groupings, this change may represent some sort of internal reorganization of intravillage corporate segments. In fact, if indeed corporate groups formed risk-sharing and food-sharing groups, then the observed variation in size and shape of jacal room blocks may have echoed differences in size and demographic composition, and economic success, of these groups. While people in the Jacal period tolerated (or encouraged) a certain level of variation in room block size, these differences are dramatically minimized in the plaza-oriented pueblos.

To understand the social significance of this architectural change, we can examine a contrasting case study from the Mesa Verde area of the northern Southwest. In the Mesa Verde region, Schachner (2010) provides an example of how the internal dynamics of social differentiation might play out in the context of early village communities at McPhee Village. In his study, he examined architectural evidence of expanding economic production to investigate intrasite differences in corporate group size and local influence. At McPhee Village, larger room blocks tend to be U-shaped, and are spatially associated with communal ritual facilities. Schachner (2010) found that the large U-shaped room room blocks were also associated with more rooms dedicated to storage, and rooms that seem to have been specialized craft production areas. In addition, the occupants of the large U-shaped room blocks appear to have engaged in more ritual activity, and to have engaged in more control over general access to ritual space (Schachner 2001). In sum, McPhee Village shows an apparent distinction between the social groups that were occupying the large U-shaped room blocks, and those occupying smaller linear ones. Groups controlling the larger room blocks within a given village site appear to represent corporate groups that were increasingly differentiating themselves not only by their control over more roofed space but also by high levels of storage, craft production, and ritual activity (Schachner 2010:490).

This association among variables of corporate group size and village subgroup control over storage facilities and ritual is not necessarily seen in other geographic areas. Schachner (2010:490) points out, for example, that the trend toward political hierarchy in the archaeological record of southeastern North America is associated more strongly with communal (village-level) storage facilities. This observation indicates that there is no necessary connection between corporate group formation and economic competition between village sub-units.

In fact, cross-cultural studies of social and organizational change suggest that in fact any organizational trend toward social differentiation may be a highly contested and uncertain process. Many ethnographic studies have shown that people in small-scale societies often resist overt expressions of difference and inequality (e.g., Wiessner 2002; Hammel 2005; Wills and Crown 2004). As Mills (2004a) also points out, some aspects of social organization in pre-state society may result in a degree of social inequality between individuals or groups, but its expression can also be minimized, masked, or even soundly rejected before it can become institutionalized (see also Wiessner 2002). Mills (2004a) examines particularly the role of objects (inalienable goods, sensu Wiener 1985) in this process within a "prestige economy"; Spielmann (2002) considers a somewhat similar situation concerning "the ritual mode of production" within Pueblo Indian society.

In small Southwestern villages, if social differences between larger and smaller room block groups (between larger and smaller corporate groups) became large or obvious enough, we might expect to see increasing resentment, resistance, and pushback on the part of the smaller groups (e.g., Emerson and Pauketat 2002). In addition, as the size of the larger corporate groups increased over time, we might also expect to see increased conflict and disagreements among individuals within these larger groups themselves (see, for example, Fowles et al. 2007). To the extent that such intergroup competition, resistance, and conflict affected the stability and viability of the village community as a whole, and to the extent that individuals and groups valued maintaining this continuity of the larger social group, we can expect that individuals and different corporate groups would need to rebalance their participation in social processes involving autonomy, competition, and cooperation in order to imagine, create, and reinforce more broadly shared social goals at this larger organizational scale.

In practical terms, I suggest that an architectural change to a standardized room block size and layout may in part represent the larger group's self-conscious realignment of social goals to emphasize similarity of any constituent social units, and to minimize intravillage social distinctions between corporate groups or other social divisions. In the Salinas region, people's early emphasis on standardization and uniformity in room block size and organization may be such a deliberate strategy of resistance to the perceived dangers of factional competition within village communities. It may be significant that the move to standardize the size and shape of room blocks occurs at the same time as a change in village layout to include a central plaza. This form of site layout not only minimizes differences both within and between room blocks by standardizing room size and arrangement but also creates a single focal point for village interaction and also social monitoring—the central plaza (see Graves and Van Keuren 2011).

In this respect, I suggest that the plaza-oriented pueblo may represent a deliberate strategy of resistance against the possible threat of increased social differentiation of village subgroups, enacted through the pervasive influence of the built environment. This architectural change may thus express the extent to which the social integration of the village as a whole community took precedence over people's alignment with and allegiance to potentially disruptive competition and conflict between subvillage social units. The rare and little-understood circular pueblo in the American Southwest may provide a similar but less adaptable strategy that people attempted to use in somewhat the same way.

The plaza-oriented layout was adopted rapidly across the Salinas region at the beginning of the Early Pueblo period, suggesting that this spatial organization was highly successful in replacing the old social rules with new ones. Plaza-oriented pueblos (and nucleated sites in general) are also effective in providing for defense actions against outsiders. However, as historic events at the Hopi village of Oraibi have demonstrated (Bradfield 1971; Krutz 1973; Cameron 1999b), there are also significant threats to village stability and continuity that originate from inside the village itself. In small-scale societies that lack well-defined supra-village institutions of conflict resolution, people rely on shared ideas to provide guidelines and standards for behavior—ideas that may find material expression in iconography (e.g., Crown 1994; Graves and Eckert 1998), objects (e.g., Spielmann 2002; Mills 2008), or in activities such as feasts (Mills 2007). In the case of early villages in Salinas, it is not clear what these new ideas may have entailed, or how they arose, or why exactly individual village communities adopted them, but the architectural result shows a temporal shift in overall social emphasis from the individual room block units to "the pueblo," the village community as a whole.

It is clear that the plaza-oriented pueblo expressed adoption of a stable and long-lasting organizational form. In fact, this larger social unit of "the plaza-oriented pueblo" then becomes the unit of reiteration in the larger sprawling Pueblo IV period pueblos, at least for a time. By the time the Spanish arrived, however, each Salinas pueblo had its own relatively coherent pueblo-level political organization, identity, and reputation. As far as we know, at this time, although there were doubtless disagreements and conflicts among individuals, households, small-scale corporate groups, and other factions, the residents identified primarily with their pueblo as a whole, and individual plaza-oriented subgroups were apparently not a particularly important organizational unit (Hayes et al. 1981; Spielmann 2002, 2004b).

Concluding Thoughts

Because of their relative rarity and also because, with incredible stupidity, I neglected to make close observation and record of the exact stratigraphic position of every specimen found, the data as to the evolution of the Pecos stone axe are not as full as could be wished.
—A.V. KIDDER, *THE ARTIFACTS OF PECOS*

The Salinas region is not particularly well-known by the archaeological standards of some portions of the American Southwest, and there is plenty of basic research that remains to be done. The research presented here summarizes much of what we do know so far about the organization and operation of early village societies in this region and also highlights those topics and questions where more information is needed.

This geographic region has many advantages for studying early village society because here, by chance, many of the different possible processes that complicate our research questions are not present. For example, because the pithouse village sites occur relatively late in time, we do not have to worry about whether people of that time period were aware of domesticated crops or the concepts involved in farming. Because we do not see the dramatic population movements and migrations that dominate the archaeological record of some other regions of the Southwest such as the Western Pueblo and the Rio Grande areas (e.g., Duff 2002; Schachner 2012), we can focus on how people created, managed, and sustained village communities under conditions in which their social institutions were not repeatedly stressed by the problems of coping with numerous immigrants. In that particular social context, we can trace through the archaeological record of the built environment how people coped with the vicissitudes of everyday life as individuals and as members of variously defined social groups.

It is particularly interesting that here, among the village societies of Salinas, the social process of constructing community apparently involved repeated and emphatic rejections of intravillage differentiation. Instead, we see people repeatedly managing the built environment as a mechanism to downplay intravillage distinctions, and to emphasize equality and unity of these social segments within the village site as a whole. I do not mean to suggest that intravillage groupings and distinctions did not exist or were not important in everyday life—only that people seem to have worked very hard to minimize the expression of any such differences within the overall built environment of the village. In light of recent research in the Mesa Verde area of the northern Southwest, in which several recent careful studies (e.g., Schachner 2010, among others) have found abundant evidence of intravillage competition, this finding is quite surprising. However, given the much smaller scale of archaeological research that has been done in Salinas, I see here only an invitation to future researchers. Much remains to be done in the Salinas Province, and in central New Mexico in general, including testing these ideas with new evidence from excavation, survey, and more detailed analyses.

References Cited

Adams, E. Charles. 1991. *The Origin and Development of the Pueblo Katsina Cult.* Tucson: University of Arizona Press.

———. 1996. "The Pueblo III–Pueblo IV Transition in the Hopi Area, Arizona." In *The Prehistoric Pueblo World, A.D. 1150–1350*, edited by Michael Adler, pp. 48–58. Tucson: University of Arizona Press.

———. 2002. *Homol'ovi: An Ancient Hopi Settlement Cluster.* Tucson: University of Arizona Press.

———, ed. 2004. *Homol'ovi IV: The First Village.* Occasional Electronic Paper 1. Arizona State Museum, University of Arizona, Tucson. Electronic document, accessed May 2010. http://www.statemuseum.arizona.edu/pubs/asmoep/001/pdf/index.shtml.

Adams, J. W. and A. B. Kasakoff. 1976. "Factors Underlying Endogamous Group Size." In *Regional Analysis Volume II: Social Systems*, edited by C.A. Smith, pp. 149–173. New York: Academic Press.

Adler, Michael A. 1989. "Ritual Facilities and Social Integration in Non-ranked Societies." In *The Architecture of Social Integration in Prehistoric Pueblos*, edited by William D. Lipe and Michelle Hegmon, pp. 35–52. Occasional Paper No. 1. Cortez: Crow Canyon Archaeological Center.

———. 1993. "Why Is a Kiva? New Interpretations of Prehistoric Social Integrative Architecture in the Northern Rio Grande Region of New Mexico." *Journal of Anthropological Research* 49:319–346.

———. 1994. "Population Aggregation and the Anasazi Social Landscape." In *The Ancient Southwestern Community*, edited by W. H. Wills and Robert D. Leonard, pp. 85–102. Albuquerque: University of New Mexico Press.

———.1996a. "Fathoming the Scale of Anasazi Communities." In *Interpreting Southwestern Diversity: Underlying Principles and Overarching Patterns*, edited by Paul R. Fish and J. Jefferson Reid, pp. 97–106. Anthropological Research Papers No. 48. Tempe: Arizona State University.

———, ed. 1996b. *The Prehistoric Pueblo World, A.D. 1150–1350.* Tucson: University of Arizona Press.

———. 2002a. "Building Consensus: Tribes, Architecture and Typology in the American Southwest." In *The Archaeology of Tribal Societies*, edited by William A. Parkinson, pp. 155–172. Archaeological Series 15. Ann Arbor: International Monographs in Prehistory.

———. 2002b. "The Ancestral Pueblo Community as Structure and Strategy." In *Seeking the Center Place: Archaeology and Ancient Communities in the Mesa Verde Region*, edited by Mark D. Varien and Richard H. Wilshusen, pp. 25–39. Salt Lake City: University of Utah Press.

———. 2002c. "Negotiating the Village: Community Landscapes in the Late Pre-Historic American Southwest." In *Inscribed Landscapes: Marking and Making Place*, edited by Bruno David and Meredith Wilson, pp. 200–216. Honolulu: University of Hawaii Press.

Adler, Michael A. and Richard Wilshusen. 1990. "Large Scale Integrative Facilities in Tribal Societies: Cross-Cultural and Southwestern U.S. Examples." *World Archaeology* 22:133–146.

Agorsah, Emmanuel Kofi. 1988. "Evaluating Spatial Behavior Patterns of Prehistoric Societies." *Journal of Anthropological Archaeology* 7:231–247.

Allen, Bruce D. 2005. "Ice Age Lakes in New Mexico." In *New Mexico's Ice Ages*, edited by S. G. Lucas, G. S. Morgan, and K. E. Zieler, pp. 107–114. Bulletin 28. Albuquerque: New Mexico Museum of Natural History and Science.

Allen, Bruce D. and R. Anderson. 2000. "A Continuous, High-Resolution Record of Late Pleistocene Climate Variability from the Estancia Basin, New Mexico." *Geological Society of America Bulletin* 112:1444–1458.

Altschul, Jeffrey H. 1978. "The Development of the Chacoan Interaction Sphere." *Journal of Anthropological Research* (1978):109–146.

Altschul, Jeffrey H. and Edgar K. Huber. 2000. "Economics, Site Structure, and Social Organization During the Basketmaker III Period." In *Foundations of Anasazi Culture: The Basketmaker-Pueblo Transition*, edited by Paul F. Reed, pp. 145–160. Salt Lake City: University of Utah Press.

Ames, Kenneth M., Cameron M. Smith, and Alexander Bourdeau. 2008. "Large Domestic Pits on the Northwest Coast of North America." *Journal of Field Archaeology* 33:3–18.

Amith, Jonathan D. 2005. "Place Making and Place Breaking: Migration and the Development Cycle of Community in Colonial Mexico." *American Ethnologist* 32:159–179.

Anderson, Benedict. 1991. *Imagined Communities: Reflections on the Origin and Spread of Nationalism*. 2nd edition. New York: Verso.

Anschuetz, Kurt F. 2006. "Tewa Fields, Tewa Traditions." In *Canyon Gardens, the Ancient Pueblo Landscapes of the American Southwest*, edited by V. B. Price and Baker H. Morrow, pp. 57–74. Albuquerque: University of New Mexico Press.

Aranyosi, E. F. 1999. "Wasteful Advertising and Variance Reduction: Darwinian Models for the Significance of Nonutilitarian Architecture." *Journal of Anthropological Archaeology* 18:356–375.

Arkush, Elizabeth N. and Mark W. Allen, eds. 2006. *The Archaeology of Warfare: Prehistories of Raiding and Conquest*. Gainesville: University Press of Florida.

Ashmore, Wendy. 1991. "Site Planning and Concepts of Directionality Among the Ancient Maya." *Latin American Antiquity* 2:199–226.

Bachhuber, F. W. 1982. "Quaternary History of the Estancia Valley, Central New Mexico." In *Albuquerque Country II*, edited by J. A. Grambling, S. G. Wells, and J. F. Callender, pp. 343–346. Guidebook No. 33. Socorro: New Mexico Geological Society.

Baker, Ele M. 1936. *Report of Stratification Tests at Quarai*. Manuscript on file, Salinas Pueblo Missions National Monument, Mountainair, New Mexico.
Baldwin, Stuart J. 1983. "A Tentative Occupational Sequence for Abó Pass, Central New Mexico." *COAS: New Mexico Archaeology and History* 1:12–28.
Bandelier, Adolphus. [1881] 2008. *Historical Introduction to Studies Among the Sedentary Indians of New Mexico: Report on the Ruins of the Pueblo of Pecos*. Papers of the Archaeological Institute of America, Vol. 1. Charleston: Bibliobazaar.
———.1890. *Final Report of Investigations Among the Indians of the Southwestern United States, Carried on Mainly in the Years from 1880 to 1885* (Part I). Cambridge: Cambridge University Press.
Bandy, Matthew S. 2004. "Fissioning, Scalar Stress, and Social Evolution in Early Village Societies." *American Anthropologist* 106:322–333.
Bandy, Matthew S. and Jake R. Fox, eds. 2010. *Becoming Villagers: Comparing Early Village Societies*. Tucson: University of Arizona Press.
Bankoff, H. Arthur and Frederick A. Winter. 1979. "A House-Burning in Serbia: What Do Burned Remains Tell an Archaeologist?" *Archaeology* 32(5):8–14.
Barnard, Alan, ed. 2004. *Hunter-Gatherers in History, Archaeology, and Anthropology*. New York: Berg Publishers.
Barrett, Eliore M. 2002. "The Geography of the Rio Grande Pueblos in the Seventeenth Century." *Ethnohistory* 49:123–169.
Barth, Fredrik. 1978. "Conclusions." In *Scale and Social Organization*, edited by Fredrik Barth, pp. 253–274. New York: Columbia University Press.
Bartlett, Mary Lee and Patricia A. McAnany. 2000. "Crafting Communities: The Materialization of Formative Mayan Identities." In *The Archaeology of Communities: A New World Perspective*, edited by Marcello. A. Canuto and Jason Yaeger, pp. 102–122. London: Routledge.
Bates, Robert L., R. H. Wilpot, J. A. MacAlpin, and George Vorbe. 1947. *The Geology of the Gran Quivira Quadrangle*. Bulletin 26. Socorro: New Mexico Bureau of Mines and Mineral Resources.
Bawden, G. 1982. "Community Organization Reflected by the Household: A Study of Pre-Columbian Social Dynamics." *Journal of Field Archaeology* 9:165–181.
Beck, Margaret E. 2006. "Midden Ceramic Assemblage Formation: A Case Study from Kalinga, Philippines." *American Antiquity* 71:27–52.
Beck, Robin A., Jr., Douglas J. Bolender, James A. Brown, and Timothy K. Earle. 2007. "Eventful Archaeology: The Place of Space in Structural Transformation." *Current Anthropology* 48:833–860.
Beckett, Patrick H. 1981. *An Archaeological Survey and Assessment of Gran Quivira National Monument, New Mexico*. National Park Service Contract Number PX7029-0-0697, Report No. 467. Cultural Resource Management Division, Department of Sociology and Anthropology, New Mexico State University, Las Cruces.
Beckett, Patrick H. and Regge Wiseman. 1982. *Laboratory of Anthropology Site Report, LA 38448*. Manuscript on file at the Museum of New Mexico, Santa Fe.
Beers, Ward. 2012. "All Along the Watchtower: Prehistoric Signalling Behavior in the Jumanos Pueblo Cluster, Torrance County, New Mexico." M.A. thesis. Department of Anthropology, Eastern New Mexico University, Portales.
Bender, Barbara. 1978. "Gatherer-Hunter to Farmer: A Social Perspective." *World Archaeology* 10:204–237.

———. 1990. "The Dynamics of Nonhierarchial Societies." In *The Evolution of Political Systems: Sociopolitics in Small-Scale Sedentary Societies*, edited by S. Upham, pp. 247–263. Cambridge: Cambridge University Press.

Bender, Donald R. 1967. "A Refinement of the Household: Families, Co-Residence and Domestic Functions." *American Anthropologist* 69(5):493–504.

Berman, Mary Jane. 1979. *Cultural Resources Overview: Socorro Area, New Mexico*. Albuquerque: U.S. Department of Agriculture, Forest Service.

Bernardini, Wesley. 1996. "Transitions in Social Organization: A Predictive Model from Southwestern Archaeology." *Journal of Anthropological Archaeology* 15:372–402.

———. 1998. "Conflict, Migration, and the Social Environment: Interpreting Architectural Change in Early and Late Pueblo IV Aggregations." In *Migration and Reorganization: The Pueblo IV Period in the American Southwest*, edited by Katherine A. Spielmann, pp. 91–114. Anthropological Research Papers No. 51. Tempe: Arizona State University.

———. 2000. "Kiln Firing Groups: Inter-Household Economic Collaboration and Social Organization in the Northern American Southwest." *American Antiquity* 65(2):365–377.

———. 2005. "Spatial and Temporal Aspects of Prehistoric Cultural Identity: A Case Study from the American Southwest." *American Antiquity* 70:31–54.

Binford, Lewis R. 1980. "'Willow Smoke and Dogs' Tails: Hunter-Gatherer Settlement Systems and Archaeological Site Formation." *American Antiquity* 45(1):4–20.

———. 1981. "Behavioral Archaeology and the 'Pompeii Premise.'" *Journal of Anthropological Research* 37(3):195–208.

———. 1982. "The Archaeology of Place." *Journal of Anthropological Archaeology* 1:5–31.

———. 1990. "Mobility, Housing, and Environment: A Comparative Study." *Journal of Anthropological Research* 46:119–152.

Blanton, Richard E. 1994. *Houses and Households: A Comparative Study*. New York: Plenum Press.

Bocquet-Appel, Jean-Pierre. 2009. "The Demographic Impact of the Agricultural System in Human History." *Current Anthropology* 50:657–660.

Bourlier, Bobby G. 1970. *Soil Survey, Torrance Area, New Mexico*. Washington, DC: U.S. Department of Agriculture.

Bowser, Brenda and John Q. Patton. 2004. "Domestic Spaces as Public Places: An Ethnoarchaeological Case Study of Houses, Gender, and Politics in the Ecuadorian Amazon." *Journal of Archaeological Method and Theory* 11:157–181.

Bowser, Brenda and María Nieves Zedeño, eds. 2009. *The Archaeology of Meaningful Places*. Salt Lake City: University of Utah Press.

Bradfield, Richard M. 1971. *The Changing Pattern of Hopi Agriculture*. Occasional Papers 30. London: Royal Anthropological Institute.

Bradfield, Wesley. 1929. *Cameron Creek Village*. Monograph 1. Santa Fe: School of American Research.

Bradley, Bruce. 1993. "Planning, Growth, and Functional Differentiation at the Prehistoric Pueblo: A Case Study from SW Colorado." *Journal of Field Archaeology* 20:23–42.

Bradley, Richard. 2012. *The Idea of Order: The Circular Archetype in Prehistoric Europe*. Oxford: Oxford University Press.

Brand, Stewart. 1995. *How Buildings Learn: What Happens After They're Built*. New York: Penguin Books.

Brandt, Elizabeth. 1995. "Egalitarianism, Hierarchy, and Centralization in the Pueblos." In *The Ancient Southwestern Community: Models and Methods of the Study of Prehistoric Social Organization*, edited by W. H. Wills and Robert D. Leonard, pp. 9–24. Albuquerque: University of New Mexico Press.

Brett, Linda C. and Phillip H. Shelley. 1985. "The Mountainair Lithic Scatters: Another Explanation for Edge Angle Distributions." In *Views of the Jornada Mogollon*, edited by Colleen M. Beck, pp. 121–129. Eastern New Mexico University Contributions in Anthropology 12. Portales: Eastern New Mexico University Press.

Brew, John O. 1946. *Archaeology of Alkali Ridge, Southeast Utah*. Papers of the Peabody Museum of Archaeology and Ethnology, Vol. 21. Cambridge: Harvard University.

Brooks, James F. 2002. *Captives and Cousins: Slavery, Kinship, and Community in the Southwest Borderlands*. Chapel Hill: University of North Carolina Press.

Brouwer, Marieka and Alison E. Rautman. 2013. *The Kite Pithouse Village (LA 38448), Final Report of the 1986 Excavations, Torrance County, New Mexico*. Manuscript on file, Museum of New Mexico, Santa Fe.

Brouwer, Marieka, Alison E. Rautman, and Matthew A. Chamberlin, eds. 2009. *Preliminary Report of the 2008 Excavations at Frank's Ruin (LA 9032), New Mexico*. Manuscript on file, New Mexico Department of Cultural Affairs and The Museum of New Mexico, Santa Fe.

Brumfiel, Elizabeth M. 1995. "Comments." In *Heterarchy and the Analysis of Complex Societies*, edited by Robert M. Ehrenreich, Carole L. Crumley, and Janet E. Levy, pp. 125–131. Archeological Papers No. 6. Washington, DC: American Anthropological Association.

Bullard, William Rotch. 1962. *The Cerro Colorado Site and Pithouse Architecture in the Southwestern United States Prior to A.D. 900*. Papers of the Peabody Museum of Archaeology and Ethnology, Vol. 44, No. 2. Cambridge: Harvard University.

Byrd, Brian F. 1994. "Public and Private, Domestic and Corporate: The Emergence of the Southwest Asian Village." *American Antiquity* 59:639–666.

Cameron, Catherine M. 1990. "The Effect of Varying Estimates of Pit Structure Use-Life on Prehistoric Population Estimates in the American Southwest." *Kiva* 55:155–166.

———. 1991. "Structure Abandonment in Villages." *Archaeological Method and Theory* 3:155–194.

———. 1999a. "Room Size, Organization of Construction, and Archaeological Interpretation in the Puebloan Southwest." *Journal of Anthropological Archaeology* 19: 201–239.

———. 1999b. *Hopi Dwellings: Architectural Change at Orayvi*. Tucson: University of Arizona Press.

Cameron, Catherine M. and Andrew I. Duff. 2008. "History and Process in Village Formation: Context and Contrasts from the Northern Southwest." *American Antiquity* 73:29–58.

Cameron, Catherine M. and Steven Tomka, eds. 1993. *Abandonment of Settlements and Regions: Ethnoarchaeological Approaches*. Cambridge: Cambridge University Press.

Cannon, Michael D. 2000. "Large Mammal Relative Abundance in Pithouse and Pueblo Period Archaeofaunas from Southwestern New Mexico: Resource Depression Among the Mimbres-Mogollon." *Journal of Anthropological Archaeology* 19: 317–347.

Canuto, Marcello A. and Jason Yaeger, eds. 2000. *The Archaeology of Communities: A New World Perspective*. New York: Routledge.

Caperton, Thomas. 1981. "An Archaeological Reconnaissance of the Gran Quivira Area." In *Contributions to Gran Quivira Archeology, Gran Quivira National Monument, New Mexico*, National Park Service, Publications in Archeology 17, edited by Alden C. Hayes, pp. 3–13. Washington, DC: U.S. Department of the Interior.

Capone, Patricia. 1995. "Mission Pueblo Ceramic Analyses: Implications for Protohistoric Interaction Networks and Cultural Dynamics." Ph.D. dissertation. Department of Anthropology, Harvard University, Cambridge.

———. 1997. *Prehistoric Craft Specialization in Non-hierarchical Societies: Petrographic Analysis of the Salinas Area, New Mexico, White Wares and Utility (Brown) Wares*. Manuscript on file, Department of Anthropology, Arizona State University, Tempe.

Carleton, James Henry. 1854. *Diary of an Excursion to the Ruins of Abó, Quarra and Gran Quivira, in New Mexico*, 9th Annual Report. Washington, DC: Smithsonian Institution Press.

Carlyle, Shawn W., Ryan L. Parr, Geoffrey Hayes, and Dennis O'Rourke. 2000. "Context of Maternal Lineages in the Greater Southwest." *American Journal of Physical Anthropology* 113:85–101.

Casey, Clyde. 2007. *The Treasure of Gran Quivira*. Bradenton: Booklocker.com.

Chamberlin, Matthew A. 2001. *Aggregation and Social Identity in Salinas: Landscape Formation from A.D. 1100–1300*. Web publication. http://archaeology.asu.edu/vm/southwest/salinas/chamberlin.html.

———. 2002. "Technology, Performance, and Intended Use: Glaze Ware Jars in the Pueblo IV Rio Grande." *Kiva* 67:269–296.

———. 2006. "Symbolic Conflict and the Spatiality of Traditions in Small-Scale Society." *Cambridge Archaeological Journal* 16:39–51.

———. 2008. "Evaluating the Origins of Complex Identities in the Ancestral Puebloan World." Ph.D. dissertation. School of Human Evolution and Social Change, Department of Anthropology, Arizona State University, Tempe.

———. 2012. "Plazas, Performance, and Symbolic Power in Ancestral Pueblo Religion." In *Religious Transformation in the Late Pre-Hispanic Pueblo World*, edited by Donna M. Glowacki and Scott Van Keuren, pp. 130–152. Tucson: University of Arizona Press.

Chamberlin, Matthew A. and Alison E. Rautman. 2009. Conflict and Its Aftermath in the Salinas Pueblo Province. Poster presented at the 74th Annual Meeting of the Society for American Archaeology, Atlanta, Georgia.

Chamberlin, Matthew A., Alison E. Rautman, and Julie Solometo. 2011. "Consolidation and Conflict." *Archaeology Southwest* 25:4–6.

Chapin, Charles E. and William R. Seager. 1975. *Evolution of the Rio Grande Rift in the Socorro and Las Cruces Areas*. Field Conference Guidebook 26. Socorro: New Mexico Geological Society.

Cheney, Fred W. 1934. "Tons of Hidden Treasure." *New Mexico Magazine* 12:16–17, 38–39, 41.

Christakis, Kostas S. 1999. "Pithoi and Food Storage in Neopalatial Crete: A Domestic Perspective." *World Archaeology* 31:1–20.

Christie, Jessica Joyce. 2009. *Landscapes of Origin in the Americas: Creation Narratives Linking Ancient Places and Present Communities*. Tuscaloosa: University of Alabama Press.

Ciolek-Torrello, Richard. 1985. "A Typology of Room Function at Grasshopper Pueblo, Arizona." *Journal of Field Archaeology* 12:41–63.

Clark, Tiffany C. 2006. "Production, Exchange, and Social Identity: A Study of Chupadero Black-on-White Pottery." Ph.D. dissertation. School of Human Evolution and Social Change, Department of Anthropology, Arizona State University, Tempe.

Cole, Sally J. 1984. *The Abó Painted Rocks Documentation and Analysis*. Salinas Pueblo Missions National Monument report. Southwestern Parks and Monuments Association.

Colton, Harold S. 1941. "Prehistoric Trade in the Southwest." *The Scientific Monthly* 52:308–319.

Coltrain, Joan Benner, Joel C. Janetski, and Shawn W. Carlyle. 2007. "The Stable- and Radio-Isotope Chemistry of Western Basketmaker Burials: Implications for Early Puebloan Diets and Origins." *American Antiquity* 72:301–321.

Colyer, Anne M. 1996. "Testing Dietary Change at the Pithouse to Pueblo Transition." M.A. thesis. Department of Anthropology, Michigan State University, East Lansing.

Corbyn, Clara A. B. 1904. *La Gran Quibira: A Musical Mystery*. Privately published by the author.

Cordell, Linda S. 1977. "Late Anasazi Farming and Hunting Strategies: One Example of a Problem in Congruence." *American Antiquity* 42(3): 449–461.

———. 1994. "Introduction: Community Dynamics of Population Aggregation in the Prehistoric Southwest." In *The Ancient Southwestern Community*, edited by W. H. Wills and Robert D. Leonard, pp. 79–83. Albuquerque: University of New Mexico Press.

———. 1995. "Tracing Migration Pathways from the Receiving End." *Journal of Anthropological Archaeology* 14:203–211.

———. 1996. "Big Sites, Big Questions: Pueblos in Transition." In *The Prehistoric Pueblo World, A.D. 1150–1350*, edited by Michael A. Adler, pp. 228–240. Tucson: University of Arizona Press.

Cordell, Linda S. and Maxine E. McBrinn. 2012. *Archaeology of the Southwest*. 3rd edition. Walnut Creek: Left Coast Press.

Cordell, Linda S., Carla R. Van West, Jeffrey S. Dean, and Deborah A. Muenchrath. 2007. "Mesa Verde Settlement History and Relocation: Climate Change, Social Networks, and Ancestral Pueblo Migration." *Kiva* 72:379–406.

Creamer, Winifred. 1996. "Developing Complexity in the American Southwest: Constructing a Model for the Rio Grande Valley." In *Emergent Complexity: The Evolution of Intermediate Societies*, edited by Jeanne E. Arnold, pp. 91–109. International Monographs in Prehistory. Ann Arbor: University of Michigan Press.

———. 2000. "Regional Interactions and Regional Systems in the Proto-Historic Rio Grande." In *The Archaeology of Regional Interaction: Religion, Warfare, and*

Exchange across the American Southwest and Beyond, edited by Michelle Hegmon, pp. 99–118. Boulder: University of Colorado Press.

Creed, Gerald W., ed. 2006. *The Seductions of Community: Emancipations, Oppressions, Quandaries*. Santa Fe: School of American Research Press.

Creel, Darrell and Roger Anyon. 2003. "New Interpretations of Mimbres Public Architecture and Space: Implications for Culture Change." *American Antiquity* 68:67–92.

Cress, Johnnie. [1957]1983. *Report on a Pueblo Refuse Pile*. Lea County Archaeological Society, Bulletin 1. Reprinted as Appendix A in *Tabira: Outpost on the East: Collected Papers in Honor of Charlie R. Steen*, edited by Nancy Fox, pp. 142–145. Papers of the Archaeological Society of New Mexico No. 8. Albuquerque: Archaeological Society of New Mexico.

Crow Canyon Archaeological Center. 2011. Peoples of the Mesa Verde Region. Available: http://www.crowcanyon.org/peoplesmesaverde.

Crown, Patricia L. 1987. "Water Storage in the Prehistoric Southwest." *Kiva* 52:209–228.

Crown, Patricia L. and Timothy A. Kohler. 1994. "Community Dynamics, Site Structure, and Aggregation in the Northern Rio Grande." In *The Ancient Southwestern Community*, edited by W. H. Wills and Robert D. Leonard, pp. 103–118. Albuquerque: University of New Mexico Press.

Culbert, James I. 1941. "Pinto Beans in the Estancia Valley of New Mexico." *Economic Geography* 17:50–60.

Cummings, Linda Scott. 1988. "Pollen Analysis for the Kite Site, New Mexico." In "The Environmental Context of Decision-Making: Coping Strategies Among Prehistoric Cultivators in Central New Mexico," Appendix G, by Alison E. Rautman. Ph.D. dissertation. Department of Anthropology, University of Michigan, University Microfilms, Ann Arbor.

———. 1995. "Pollen Analysis." In *Final Report on the Excavation of Pueblo de la Mesa, Cibola National Forest, Torrance County, New Mexico*, by Alison E. Rautman, pp. 125–139. Manuscript on file. USDA Forest Service, Cibola National Forest, Albuquerque, New Mexico.

Cummings, Linda Scott, Kathryn Puseman, and Thomas E. Moutoux. 1997. *Pollen and Macrofloral Analysis at Kite Pueblo, Site LA 199, New Mexico*. Technical Report 96–41. Paleo Research Laboratories, Denver, Colorado.

Cushing, Frank Hamilton. 1901. *Zuni Folk Tales*. London: G. P. Putnam's Sons.

Cutting, Marion. 2003. "The Use of Spatial Analysis to Study Prehistoric Settlement Architecture." *Oxford Journal of Archaeology* 22:1–21.

Darling, Andrew. 1999. "Mass Inhumation and the Execution of Witches in the American Southwest." *American Anthropologist* 100:732–752.

David, Nicholas. 1971. "The Fulani Compound and the Archaeologist." *World Archaeology* 3:111–131.

Dean, Jeffrey S. 1996. "Kayenta Anasazi Settlement Transformations in Northeastern Arizona: AD 1150 to 1350." In *The Prehistoric Pueblo World A.D. 1150–1350*, edited by Michael A. Adler, pp. 29–47. Tucson: University of Arizona Press.

DeBoer, Warren R. 1988. "Subterranean Storage and the Organization of Surplus: The View from Eastern North America." *Southeastern Archaeology* 7:1–20.

Diehl, Michael. 1997. "Changes in Architecture and Land Use Strategies in the American Southwest: Upland Mogollon Pithouse Dwellers, A.C. 200–1000." *Journal of Field Archaeology* 24:179–194.
Diehl, Michael and Steven A. LeBlanc, eds. 2001. *Early Pithouse Villages of the Mimbres Mogollon and Their Regional Context.* Peabody Museum of Archaeology and Ethnology, Harvard University. Cambridge: Peabody Museum Press.
Dohm, Karin. 1990. "Effect of Population Nucleation on House Size for Pueblos in the American Southwest." *Journal of Anthropological Archaeology* 9:201–239.
———. 1994. "The Search for Anasazi Village Origins: Basketmaker II Dwelling Aggregation on Cedar Mesa." *Kiva* 60:257–276.
Dominguez, Steven and Kenneth E. Kolm. 2005. "Beyond Water Harvesting: A Soil Hydrology Perspective on Traditional Southwestern Agricultural Technology." *American Antiquity* 70:732–765.
Donham, Donald L. 1981. "Beyond the Domestic Mode of Production." *Man* 16:515–541.
Doolittle, William E. 2004. "Gardens Are Us, We Are Nature: Transcending Antiquity and Modernity." *Geographical Review* 94:391–404.
Doolittle, William E. and Jonathan B. Mabry. 2006. "Environmental Mosaics, Agricultural Diversity, and the Evolutionary Adoption of Maize in the American Southwest." In *Histories of Maize: Multidisciplinary Approaches to the Prehistory, Linguistics, Biogeography, Domestication, and Evolution of Maize,* edited by John E. Staller, Robert H. Tykot, and Bruce F. Benz, pp 109–122. New York: Academic Press.
Dovey, Kim. 1999. *Framing Places: Mediating Power in Built Form.* London: Routledge.
Doxiadis, Constantinos. 1968. *Ekistics: An Introduction to the Science of Human Settlements.* Oxford: Oxford University Press.
Dozier, Edward P. 1965. "Southwestern Social Units and Archaeology." *American Antiquity* 31:38–47.
———. 1966. *Hano, A Tewa Indian Community in Arizona.* New York: Holt, Rinehart and Winston.
Draper, Patricia. 1974. "Crowding Among Hunter-Gatherers: The !Kung Bushmen." *Science* 182(4019):301–303.
Duff, Andrew. 1998. "The Process of Migration in the Late Prehistoric Southwest." In *Migration and Reorganization: The Pueblo IV Period in the American Southwest,* edited by Katherine A. Spielmann, pp. 31–52. Arizona State University Anthropological Research Paper 51. Tempe: Arizona State University.
———. 2002. *Western Pueblo Identities: Regional Interaction, Migration, and Transformation.* Tucson: University of Arizona Press.
———. 2004. "Settlement Clustering and Village Interaction in the Upper Little Colorado Region." In *The Protohistoric Pueblo World, A.D. 1275–1600,* edited by E. Charles Adams and Andrew Duff, pp. 75–84. Tucson: University of Arizona Press.
———. 2005. "On the Fringe: Community Dynamics at Cox Ranch Pueblo." In *Proceedings of the 13th Mogollon Archaeology Conference,* edited by Lonnie Ludeman, pp. 303–319. Las Cruces: New Mexico State University.
Dunnell, Robert C. 1999. "The Concept of Waste in an Evolutionary Archaeology." *Journal of Anthropological Archaeology* 18:243–250.

Durkheim, Emile. [1893]1933. *The Division of Labor in Society*. Translated by George Simpson. New York: The Free Press.

Eckert, Suzanne L. 2006a. "Black-on-White to Glaze-on-Red: The Adoption of Glaze Technology in the Central Rio Grande Valley." In *The Social Life of Pots: Glaze Wares and Cultural Dynamics in the Southwest, A.D. 1250–1680*, edited by Judith A. Habicht-Mauche, Suzanne L. Eckert, and Deborah L. Huntley, pp. 163–178. Tucson: The University of Arizona Press.

———. 2006b. "The Production and Distribution of Glaze Painted Pottery in the Pueblo Southwest: A Synthesis." In *The Social Life of Pots: Glaze Wares and Cultural Dynamics in the Southwest, A.D. 1250–1680*, edited by Judith A. Habicht-Mauche, Suzanne L. Eckert, and Deborah L. Huntley, pp. 34–59. Tucson: University of Arizona Press.

———. 2008. *Pottery and Practice: The Expression of Identity at Pottery Mound and Hummingbird Pueblo*. Albuquerque: University of New Mexico Press.

Eco, Umberto. 1980. "Function and Sign: The Semiotics of Architecture." In *Signs, Symbols and Architecture*, edited by G. Broadbent, R. Bunt, and C. Jenks, pp. 11–69. Chichester: John Wiley and Sons.

Elsinger, Laura and Holmes Semken. 1990. "Paleoecology of Micromammals from Gran Quivira Pueblo and the Kite Site." In "The Environmental Context of Decision-Making: Coping Strategies Among Prehistoric Cultivators in Central New Mexico," Appendix F, by Alison E. Rautman. Ph.D. dissertation. Department of Anthropology, University of Michigan. University Microfilms, Ann Arbor.

Ely, Albert Grim. 1935. "The Excavation and Repair of the Quarai Mission." *El Palacio* 39:133–144.

Emerson, Thomas E. and Timothy Pauketat. 2002. "Embodying Power and Resistance at Cahokia." In *The Dynamics of Power*, edited by Maria O'Donovan, pp. 105–125. Center for Archaeological Investigations, Occasional Paper No. 30, Southern Illinois University, Carbondale.

Evans, Glen L. 1951. "Prehistoric Wells in Eastern New Mexico." *American Antiquity* 17:1–9

Ewart, Elizabeth. 2003. "Lines and Circles: Images of Time in a Panará Village." *The Journal of the Royal Anthropological Institute* 9:261–279.

Faith, J. Tyler and Adam D. Gordon. 2007. "Skeletal Element Abundances in Archaeofaunal Assemblages: Economic Utility, Sample Size, and Assessment of Carcass Transport Strategies." *Journal of Archaeological Science* 34: 872–882.

Fehr, Ernst and Klaus Schmidt. 1999. "A Theory of Fairness, Competition, and Cooperation." *The Quarterly Journal of Economics* 114(3):817–868.

Feinman, Gary M. 2000. "Corporate/Network: New Perspectives on Models of Political Action and the Puebloan Southwest." In *Social Theory in Archaeology*, edited by Michael B. Schiffer, pp. 31–51. Salt Lake City: University of Utah Press.

Feinman, Gary M., Kent G. Lightfoot, and Steadman Upham. 2000. "Political Hierarchies and Organizational Strategies in the Puebloan Southwest." *American Antiquity* 65:449–470.

Fenenga, Franklin. 1956. "Excavations at Site LA 2579, a Mogollon Village near Gran Quivira, New Mexico." In *Pipeline Archaeology: Reports of Salvage Operations in the Southwest on El Paso Natural Gas Company Projects, 1950–1953*, edited by

F. Wendorf, N. Fox, and O. L. Lewis, pp. 226–233. Flagstaff: Museum of Northern Arizona.

Ferdon, E. N. and E. K. Reed. 1950. "A Pithouse Site near Belen, New Mexico." *El Palacio* 57:40–41.

Ferguson, T. J. 1989. "Comment on Integration and Architecture." In *The Architecture of Social Integration in Prehistoric Pueblos,* edited by William D. Lipe and Michelle Hegmon, pp.169–174. Occasional Paper 1. Cortez: Crow Canyon Archaeological Center.

Fewkes, Jesse Walter. 1904. *Two Summers' Work in Pueblo Ruins.* Bureau of American Ethnology, 22nd Annual Report. Washington, DC: U.S. Government Printing Office.

Fish, Suzanne K. and Paul R. Fish, eds. 1984. *Prehistoric Agricultural Strategies in the Southwest.* Anthropological Research Papers No. 33. Tempe: Arizona State University.

Fisher, Kevin. 2009. "Placing Social Interaction: An Integrative Approach to Analyzing Past Built Environments." *Journal of Anthropological Archaeology* 28:439–457.

Fiske, Alan Page. 1994. "Relativity Within Mosse ("Mossi") Culture: Four Incommensurable Models for Social Relationships." *Ethos* 18:180–204.

Flanagan, James G. 1989. "Hierarchy in Simple 'Egalitarian' Societies." *Annual Review of Anthropology* 18:245–266.

Flannery, Kent. 1972. "The Origin of the Village as a Settlement Type in Mesoamerica and the Near East: A Comparative Approach." In *Man, Settlement and Urbanism,* edited by Peter J. Ucko, Ruth Tringham, and George W. Dimbleby, pp. 23–53. Cambridge: Gerald Duckworth.

———. 2002. "The Origins of the Village Revisited: From Nuclear to Extended Households." *American Antiquity* 67:417–433.

Flannery, Kent V. and Joyce Marcus. 2012. *The Creation of Inequality: How Our Prehistoric Ancestors Set the Stage for Monarchy, Slavery, and Empire.* Cambridge: Harvard University Press.

Flint, Richard and Shirley Cushing Flint, eds. 2005. *Documents of the Coronado Expedition, 1539–1542: "They Were Not Familiar with His Majesty, nor Did They Wish to Be His Subjects."* Dallas: Southern Methodist University.

Ford, Richard I., Albert H. Schroeder, and Stewart L. Peckham. 1972. "Three Perspectives on Puebloan Prehistory." In *New Perspectives on the Pueblos,* edited by Alfonzo Ortiz, pp. 19–39. Albuquerque: University of New Mexico Press.

Forrest, Earle Robert. 1929. *Missions and Pueblos of the Old Southwest.* Cleveland: The Arthur H. Clark Company.

Foss, Jacob. 2009. "A Comparison of the Organization of Lithic Technology at Two Prehistoric Pueblos in Central New Mexico." M.A. thesis. Department of Anthropology. Michigan State University, East Lansing.

Foster, George M. 1960. *Culture and Conquest: America's Spanish Heritage.* Viking Fund Publications in Anthropology 27. New York: Wenner-Gren Foundation for Anthropological Research Inc.

Foucault, Michel. 1979. *Discipline and Punish: The Birth of the Prison.* New York: Vintage Press.

Fowles, Severin M. 2009. "The Enshrined Pueblo: Villagescape and Cosmos in the Northern Rio Grande." *American Antiquity* 74:448–466.

Fowles, Severin M., Leah Minc, Samuel Duwe, and David V. Hill. 2007. "Clay, Conflict, and Village Aggregation: Compositional Analyses of Preclassic Pottery from Taos, New Mexico." *American Antiquity* 72:125–152.

Gaines, Sylvia W. and Warren M. Gaines. 2000. "Impact of Small-Group Decision Making in Reducing Stress Conditions." *Journal of Anthropological Archaeology* 19:103–130.

Geib, Phil R. and Kimberly Spurr. 2000. "The Basketmaker II–III Transition on the Rainbow Plateau." In *Foundations of Anasazi Culture: The Basketmaker-Pueblo Transition*, edited by Paul F. Reed, pp. 15–202. Salt Lake City: University of Utah Press.

Gilman, Patricia A. 1987. "Architecture as Artifact: Pit Structures and Pueblos in the American Southwest." *American Antiquity* 52:538–564.

———. 1995. "Multiple Dimensions of the Archaic-to-Pit Structure Period Transition in Southeastern Arizona." *The Kiva* 60(3):538–564.

———. 1997. *Wandering Villagers: Pitstructures, Mobility and Agriculture in Southeastern Arizona*. Anthropological Research Papers No. 49. Tempe: Arizona State University.

Gilman, Patricia A. and Lisa C. Young. 2005. *The Pithouse and the Pendulum: Changing Views of Pithouse Communities in the North American Southwest*. Paper presented at the Society for American Archaeology meeting, Salt Lake City, UT.

Gilpin, Dennis. 2003. "Chaco-Era Site Clustering and the Concept of Communities." *Kiva* 69:171–206.

Glassow, Michael A. 1972. "Changes in the Adaptations of Southwestern Basketmakers: A Systems Perspective." In *Contemporary Archaeology: A Guide to Theory and Contributions*, edited by Mark P. Leone, pp. 289–302. Carbondale: Southern Illinois University Press.

Glowacki, Donna M. 2010. "The Social and Cultural Contexts of the Central Mesa Verde Region During the Thirteenth-Century Migrations." In *Leaving Mesa Verde: Peril and Change in the Thirteenth-Century Southwest*, edited by Timothy A. Kohler, Mark D. Varien, and Aaron M. Wright, pp. 200–221. Tucson: University of Arizona Press.

Goffman, Erving. 1971. *Relations in Public*. New York: Basic Books.

Goody, Jack, ed. 1971. *The Developmental Cycle in Domestic Groups*. Volume 1 of Cambridge Papers in Social Anthropology. Cambridge: Cambridge University Press.

Graesch, Anthony P. 2004. "Notions of Family Embedded in the House." *Anthropology News* May 45(5):20.

Graves, William A. 2002. "Power, Autonomy, and Inequality in Rio Grande Puebloan Society, A.D. 1300–1672." Ph.D. dissertation. School of Human Evolution and Social Change, Department of Anthropology, Arizona State University, Tempe.

———. 2005. Field Notes from the 2005 Archaeological Excavations at Abó (LA 97). Salinas Pueblo Missions National Monument, Mountainair, New Mexico.

———. 2011. "Social Identity and the Internal Organization of Settlement Clusters in the Salinas District, Central New Mexico." In *The Protohistoric Pueblo World, A.D. 1275–1540*, edited by E. Charles Adams and Andrew Duff, pp. 43–52. Tucson: University of Arizona Press.

Graves, William M. and Suzanne L. Eckert. 1998. "Decorated Ceramic Distributions and Ideological Developments in the Northern and Central Rio Grande Valley,

New Mexico." In *Migration and Reorganization: The Pueblo IV Period in the American Southwest*, edited by K. A. Spielmann, pp. 263–283. Anthropological Research Papers No. 51. Tempe: Arizona State University.

Graves, William M. and Katherine A. Spielmann. 1999. *Report of the 1999 Excavations at Pueblo Blanco (LA 51), New Mexico*. Manuscript on file, USDA Forest Service, Albuquerque, New Mexico.

———. 2000a. *Report of the 2000 Excavations at Pueblo Blanco (LA 51), New Mexico*. Manuscript on file, USDA Forest Service, Albuquerque, New Mexico.

———.2000b. "Leadership, Long-Distance Exchange, and Feasting in the Protohistoric Rio Grande." In *Alternative Leadership Strategies in the Prehispanic Southwest*, edited by Barbara J. Mills, pp. 45–59. Tucson: University of Arizona Press.

Graves, William M. and Scott Van Keuren. 2011. "Ancestral Pueblo Villages and the Panoptic Gaze of the Commune." *Cambridge Archaeological Journal* 21:263–282.

Green, Earl. 1955. "Excavations near Gran Quivira, New Mexico." *Bulletin of the Texas Archaeological Society* 26:182–185.

Gregg, Josiah. [1844]1967. *Commerce of the Prairies: The Journal of a Santa Fe Trader*. Lincoln: University of Nebraska Press.

Gurven, Michael, Monique Borgerhoff Mulder, Paul L. Hooper, Hillard Kaplan, Robert Quinlan, Rebecca Sear, Sric Schniter, Christopher von Rueden, Samuel Bowles, Tom Hertz, and Adrian Bell. 2010. "Domestication Alone Does Not Lead to Inequality." *Current Anthropology* 51:49–64.

Habicht-Mauche, Judith A. 1993. *The Pottery from Arroyo Hondo Pueblo, New Mexico: Tribalization and Trade in the Northern Rio Grande*. Arroyo Hondo Archaeological Series 8. Santa Fe: School of American Research Press.

———. 2006. "The Social History of Southwestern Glaze Wares." In *The Social Life of Pots: Glaze Wares and Cultural Dynamics in the Southwest, A.D. 1250–1680*, edited by Judith A. Habicht-Mauche, Suzanne L. Eckert, and Deborah L. Huntley, pp. 3–16. Tucson: University of Arizona Press.

Habicht-Mauche, Judith A., Suzanne L. Eckert, and Deborah L. Huntley, eds. 2006. *The Social Life of Pots: Glaze Wares and Cultural Dynamics in the Southwest, A.D. 1250–1680*. Tucson: University of Arizona Press.

Hackett, Charles W. 1937. *Historical Documents Relating to New Mexico, Nueva Vizcaya, and Approaches Thereto, to 1773*. Issue 330, Volume 3. Washington, DC: Carnegie Institution of Washington.

Hall, Edward T. 1966. *The Hidden Dimension*. New York: Doubleday.

Hammel, E. A. 2005. "Kinship-Based Politics and the Optimal Size of Kin Groups." *Proceedings of the National Academy of Sciences* 102:11951–11956.

Hammond, George P. and Agapito Rey. 1940. *Narratives of the Coronado Expedition*. Albuquerque: University of New Mexico Press.

———.1966. *The Rediscovery of New Mexico, 1580–1594*. Albuquerque: University of New Mexico Press.

Hanson, J. and B. Hillier. 1982. "Domestic Space Organization." *Architectural Behavior* 2: 5–25.

Hard, Robert J., Raymond P. Mauldin, and Gerry R. Raymond. 1996. "Mano Size, Stable Carbon Isotope Ratios, and Macrobotanical Remains as Multiple Lines of Evidence of Maize Dependence in the American Southwest." *Journal of Archaeological Method and Theory* 3:253–318.

Haury, Emily W. 1992. "Thoughts After Sixty Years as a Southwestern Archaeologist." In *Emily W. Haury's Prehistory of the American Southwest,* edited by J. Jefferson Reid and David E. Doyel, pp. 435–464. Tucson: University of Arizona Press.

Haviland, William A. 1988. "Musical Hammocks at Tikal: Problems with Reconstructing Household Composition." In *Household and Community in the Mesoamerican Past,* edited by Richard Wilk and Wendy Ashmore, pp. 121–134. Albuquerque: University of New Mexico Press.

Hayden, Brian. 1995. "Pathways to Power: Principles for Creating Socioeconomic Inequalities." In *Foundations of Social Inequality,* edited by T. Douglas Price and Gary M. Feinman, pp. 15–86. New York: Plenum Press.

———. 2000. "On Territoriality and Sedentism." *Current Anthropology* 41:109–112.

Hayes, Alden C., ed. 1981. "Introduction." In *Contributions to Gran Quivira Archeology, Gran Quivira National Monument, New Mexico,* pp. 1–3. National Park Service Publications in Archeology 17. Washington, DC: U.S. Department of the Interior.

———. 1993. "The Jumanos Pueblos." In *Salinas: Archaeology, History, and Prehistory,* edited by David Grant Noble, pp. 11–15. Santa Fe: Ancient City Press.

Hayes, Alden C., J. N. Young, and A. H. Warren, eds. 1981. *Excavation of Mound 7.* National Park Service Publications in Archeology 16. Washington, DC: U.S. Department of the Interior.

Haynes, C. Vance. 1955. "Evidence of Early Man in Torrance County, New Mexico." *Bulletin of the Texas Archeological Society* 26:144–164.

Hegmon, Michelle. 1989. "Social Integration and Architecture." In *The Architecture of Social Integration,* edited by William D. Lipe and Michelle Hegmon, pp. 5–14. Occasional Paper No.1. Cortez: Crow Canyon Archaeological Center.

———. 1996. "Variability in Food Production, Strategies of Storage and Sharing, and the Pithouse-to-Pueblo Transition in the Northern Southwest." In *Evolving Complexity and Environmental Risk in Prehistoric Southwest,* edited by Joseph A. Tainter and Bonnie Bagley Tainter, pp. 223–250. Santa Fe Institute Studies in the Sciences of Complexity, Vol. 24. Reading: Addison-Wesley.

Hegmon, Michelle, Scott G. Ortman, and Jeannette L. Mobley-Tanaka. 2000. "Women, Men, and the Organization of Space." In *Women & Men in the Prehispanic Southwest: Labor, Power, and Prestige,* edited by Patricia L. Crown, pp. 43–90. Advanced Seminar Series, School of American Research. Santa Fe: School of American Research Press.

Hendon, Julia A. 1996. "Archaeological Approaches to the Organization of Domestic Labor: Household Practice and Domestic Relations." *Annual Review of Anthropology* 25:45–61.

Herrman, Bert. 2003. *Mountainair, New Mexico Centennial History 1903–2003.* Mountainair: Mountainair Public Schools.

Heyman, Marjorie W. 2004. *Using Ground Penetrating Radar: Results of the Pilot Study at Two Pithouse Settlements, Salinas, New Mexico.* Manuscript submitted to the National Park Service, Salinas Pueblo Missions National Monument, Mountainair, New Mexico.

Hickerson, Nancy P. 1994. *The Jumanos: Hunters and Traders of the South Plains.* Austin: University of Texas Press.

Hill, J. Brett. 1998. "Ecological Variability and Agricultural Specialization Among the Protohistoric Pueblos of Central New Mexico." *Journal of Field Archaeology* 25:275–294.

Hill, James. 1982. *Broken K Pueblo: Prehistoric Social Organization in the American Southwest*. Anthropological Papers No. 18. Tucson: University of Arizona Press.

Hill, W. W. 1938. *The Agricultural and Hunting Methods of the Navaho Indians*. Yale University Publications in Anthropology 18. New Haven: Yale University Press.

Hillier, Bill and Julienne Hanson. 1984. *The Social Logic of Space*. London: Cambridge University Press.

Holliday, Vance T., Bruce B. Huckell, and Robert H. Weber. 2007. "Investigations at the Mockingbird Gap Clovis Site." Department of Anthropology, University of Arizona. Accessed August 2008. http://www.argonaut.arizona.edu/projects/mockingbirdgap.htm.

Howard, Richard M. 1959. "Comments on the Indians' Water Supply at Gran Quivira National Monument." *El Palacio* 66(3):85–91.

———. 1981. "An Adobe-Lined Pit at Gran Quivira." In *Contributions to Gran Quivira Archeology, Gran Quivira National Monument, New Mexico*, edited by Alden C. Hayes, pp. 13–15. National Park Service, Publications in Archeology 17. Washington, DC: U.S. Department of the Interior.

Huckell, B. H., V. T. Holliday, R. H. Weber, and J. H. Mayer. 2006. "Archaeological and Geological Test Investigations at the Mockingbird Gap Clovis Site, Central New Mexico." *Current Research in the Pleistocene* 23:115–116.

Huntley, Deborah L., and Keith W. Kintigh. 2004. "Archaeological Patterning and Organizational Scale of Late Prehistoric Settlement Clusters in the Zuni Region of New Mexico." In *The Protohistoric Pueblo World, A.D. 1275–1600*, edited by E. Charles Adams and Andrew Duff, pp. 75–84. Tucson: University of Arizona Press.

Huntley, Deborah L., Katherine A. Spielmann, Judith A. Habicht-Mauche, Cynthia L. Herhahn, and A. Russell Flegal. 2007. "Local Recipes or Distant Commodities? Lead Isotope and Chemical Compositional Analysis of Glaze Paints from the Salinas Pueblos, New Mexico." *Journal of Archaeological Science* 34:1135–1147.

Hurt, Wesley R. 1990. *The 1939–1940 Excavation Project at Quarai Pueblo and Mission Buildings: Salinas Pueblo Missions National Monument, New Mexico*. Professional Papers 29. Santa Fe: History Division, Southwest Cultural Resources Center.

Hyland, Stanley E., ed. 2005. *Community Building in the Twenty-First Century*. Santa Fe: School of American Research Press.

Ice, Ronald. 1968. A *Report on the 1964 Excavations at Gran Quivira*. Manuscript on file, Salinas Pueblo Missions National Monument, New Mexico.

Ingold, Tim. 2007. *Lines: A Brief History*. London: Routledge.

Inomata, Takeshi. 2006. "Plaza, Performers, and Spectators: Political Theaters of the Classic Maya." *Current Anthropology* 47:805–842.

Irwin-Williams, Cynthia. 1979. "Post-Pleistocene Archaeology, 7000–2000 B.C." In *Handbook of North American Indians, volume 9*, edited by Alfonso A. Ortiz, pp. 31–42. Washington, DC: Smithsonian Institution Press.

Isbell, William H. 2000. "What We Should Be Studying: The 'Imagined Community' and the 'Natural Community.'" In *The Archaeology of Communities: A New World*

Perspective, edited by Marcello A. Canuto and Jason Yaeger, pp. 243–266. New York: Routledge.

Ivey, James E. 1988. *In the Midst of a Loneliness: The Architectural History of the Salinas Missions*. Southwest Cultural Resources Center, Professional Papers 15. Santa Fe: Southwest Regional Office, National Park Service.

James, Steven R. 1997. "Change and Continuity in Western Pueblo Households During the Historic Period in the American Southwest." *World Archaeology* 28:429–456.

Johnson, Gregory A. 1982. "Organizational Structure and Scalar Stress." In *Theory and Explanation in Archaeology*, edited by Colin Renfrew, M. J. Rowlands, and Barbara A. Seagraves, pp 389–421. New York: Academic Press.

———. 1989. "Far Outside, Looking In." In *Dynamics of Southwest Prehistory*, edited by George Gumerman and Linda Cordell, pp. 371–389. Washington, DC: Smithsonian Institution Press.

Johnson, Matthew. 1999. *Archaeological Theory: An Introduction*. Oxford: Blackwell Publishers.

Jones, Andrew. 2002. *Archaeological Theory and Scientific Practice*. Cambridge: Cambridge University Press.

Jones, D. Gareth and Robyn J. Harris. 1998. "Archeological Human Remains: Scientific, Cultural, and Ethical Considerations." *Current Anthropology* 39:253–264.

Judd, Neil M. 1916. *The Use of Adobe in Prehistoric Dwellings of the Southwest*. Washington, DC: National Museum.

Judge, W. J. 1973. *The Paleo-Indian Occupation of the Central Rio Grande Valley*. Albuquerque: University of New Mexico Press.

Kahldahl, Eric J., Scott Van Keuren, and Barbara Mills. 2004. "Migration, Factionalism, and the Trajectories of Pueblo IV Period Clusters in the Mogollon Rim Region." In *The Protohistoric Pueblo World, A.D. 1275–1600*, edited by Charles Adams and Andrew D. Duff, pp. 85–94. Tucson: University of Arizona Press.

Kantner, John. 1996. "Political Competition Among the Chaco Anasazi of the American Southwest." *Journal of Anthropological Archaeology* 15:41–105.

———. 2004. *The Ancient Puebloan Southwest*. Cambridge: Cambridge University Press.

Katzenberg, Anna and Jane H. Kelley. 1991. "Stable Isotope Analysis of Prehistoric Bone from the Sierra Blanca Region of New Mexico." In *Mogollon V*, edited by Patrick H. Beckett, pp. 207–219. Las Cruces: COAS Publishing and Research.

Keeley, L. H. 1996. *War Before Civilization*. New York: Oxford University Press.

Kelley, Jane Holden. 1984. *The Archaeology of the Sierra Blanca Region of Southeastern New Mexico*. Anthropological Papers 74. Ann Arbor: Museum of Anthropology, University of Michigan.

———. 1991. "An Overview of the Capitan North Project." In *Mogollon V*, edited by Patrick H. Beckett, pp. 166–178. Las Cruces: COAS Publishing and Research.

Kelly, Raymond C. 2000. *Warless Societies and the Origins of War*. Ann Arbor: University of Michigan Press.

———. 2005. "The Evolution of Lethal Intergroup Violence." *Proceedings of the National Academy of Sciences* 102(43):15294–15298.

Kelly, Robert L. 1992. "Mobility/Sedentism; Concepts, Archaeological Measures, and Effects." *Annual Review of Anthropology* 21:43–66.

———. 1995. *The Foraging Spectrum: Diversity in Hunter-Gatherer Lifeways*. Washington, DC: Smithsonian Institution Press.

Kent, Susan. 1984. *Analyzing Activity Areas: An Ethnoarchaeological Study of the Use of Space*. Albuquerque: University of New Mexico Press.

———, ed. 1990. *Domestic Architecture and the Use of Space: An Interdisciplinary Cross-Cultural Study*. New York: Cambridge University Press.

———. 1991. "The Relationship Between Mobility Strategies and Site Structure." In *The Interpretation of Archaeological Spatial Patterning*, edited by Ellen M. Kroll and T. Douglas Price, pp. 33–60. New York: Plenum Press.

———. 1992. "Studying Variability in the Archeological Record: An Ethnoarchaeological Model for Distinguishing Mobility Patterns." *American Antiquity* 57:635–660.

———. 1999. "The Archaeological Visibility of Storage: Delineating Storage from Trash Areas." *American Antiquity* 64:79–94.

Kenyon, John R. 1990. *Medieval Fortifications*. London: Continuum.

Kenzle, Susan C. 1997. "Enclosing Walls in the Northern San Juan: Sociophysical Boundaries and Defensive Fortifications in the American Southwest." *Journal of Field Archaeology* 24:195–210.

Kessell, John L. 1979. *Kiva, Cross, and Crown: The Pecos Indians and New Mexico, 1540–1840*. Albuquerque: University of New Mexico Press.

———. 2008. *Pueblos, Spaniards, and the Kingdom of New Mexico*. Norman: University of Oklahoma Press.

Kidder, Alfred Vincent. [1924]1968. *An Introduction to the Study of Southwestern Archaeology*. New Haven: Yale University Press.

———. 1932. *The Artifacts of Pecos*. Robert S. Peabody Foundation for Archaeology. New Haven: Yale University Press.

———. [1949]1976. "Introduction." In *Prehistoric Southwesterners from Basketmaker to Pueblo*, by Charles Avery Amsden, pp. xi–xiv. Los Angeles: Southwest Museum.

Kidder, Tristram R. 2004. "Plaza as Architecture: An Example from the Raffman Site, Northeast Louisiana." *American Antiquity* 69:514–532.

Kintigh, Keith W. 1994. "Chaco, Communal Architecture, and Cibolan Aggregation." In *The Ancient Southwestern Community*, edited by W. H. Wills and Robert D. Leonard, pp. 131–140. Albuquerque: University of New Mexico Press.

Kohler, Timothy A. 1992. "Field Houses, Villages, and the Tragedy of the Commons in the Early Northern Anasazi Southwest." *American Antiquity* 57:617–635.

Kohler, Timothy A., Matt Pier Glaude, Jean-Pierre Bocquet-Appel, and Brian M. Kemp. 2008. "The Neolithic Demographic Transition in the U.S. Southwest." *American Antiquity* 73:645–669.

Kohler, Timothy A. and Lynne Sebastian. 1996. "Population Aggregation in the Prehistoric North American Southwest." *American Antiquity* 61:597–602.

Kohler, Timothy A., Stephanie Van Buskirk, and Samantha Ruscavage-Barz. 2004. "Vessels and Villages: Evidence for Conformist Transmission in Early Village Aggregations on the Pajarito Plateau, New Mexico." *Journal of Anthropological Archaeology* 23:100–118.

Kohler, Timothy A. and Mark D. Varien, eds. 2012. *Emergence and Collapse of Early Villages: Models of Central Mesa Verde Archaeology*. Berkeley: University of California Press.

Kohler, Timothy A., Mark D. Varien, and Aaron M. Wright, eds. 2010. *Leaving Mesa Verde: Peril and Change in the Thirteenth-Century Southwest.* Tucson: University of Arizona Press.

Kolb, Charles C., Thomas H. Charlton, Warren DeBoer, Roland Fletcher, Paul F. Healy, Robert R. Janes, Raoul Naroll, and Daniel Shea. 1985. "Demographic Estimates in Archaeology: Contributions from Ethnoarchaeology on Mesoamerican Peasants." *Current Anthropology* 26:581–599.

Kolb, Michael J. and James E. Snead. 1997. "It's a Small World After All: Comparative Analyses of Community Organization in Archaeology." *American Antiquity* 62:609–628.

Kosse, K. 1994. "The Evolution of Large, Complex Groups: A Hypothesis." *Journal of Anthropological Archaeology* 13:35–50.

Kuijt, Ian. 2000. "People and Space in Early Agricultural Villages: Exploring Daily Lives, Community Size, and Architecture in the Late Pre-Pottery Neolithic." *Journal of Anthropological Archaeology* 19:75–102.

———. 2002. *Life in Neolithic Farming Communities: Social Organization, Identity, and Differentiation.* New York: Springer Verlag.

Kyte, Michael. 1988. "A Ceramic Sequence from the Chupadera Arroyo Basin, Central New Mexico." M.A. thesis. Department of Anthropology, Eastern New Mexico University, Portales.

———. 1989. "Pottery of the Project Area." In *Archaeological Reconnaissance of the Chupadera Arroyo Drainage, Central New Mexico*, edited by John Montgomery and Kathleen Bowman, pp. 123–142. Portales: Eastern New Mexico University.

Lahiji, Nadir, ed. 2011. *The Political Unconscious of Architecture: Re-opening Jameson's Narrative.* Burlington: Ashgate Publishing Company.

Lang, Richard and A. H. Harris. 1984. *The Faunal Remains from Arroyo Hondo Pueblo, New Mexico, a Short-Term Study in Subsistence Change.* Arroyo Hondo Archaeological Series 5. Santa Fe: School of American Research Press.

Lange, Charles H. 1959. *Cochiti: A New Mexico Pueblo, Past and Present.* Austin: University of Texas Press.

Lange, Charles H. and Carroll L. Riley. 1966. *The Southwestern Journals of Adolph Bandelier, 1880–1882.* Albuquerque: University of New Mexico Press.

Lavender, Catherine. 2006. *Scientists and Storytellers: Feminist Anthropologists and the Construction of the American Southwest.* Albuquerque: University of New Mexico Press.

Lawrence, Denise L. and Setha M. Low. 1990. "The Built Environment and Spatial Form." *Annual Review of Anthropology* 19:453–505.

Lawrence, Roderick J. 1996. "The Multi-dimensional Nature of Boundaries: An Integrative Historical Perspective." In *Setting Boundaries: The Anthropology of Spatial and Social Organization*, edited by Deborah Pellow, pp. 9–36. Westport: Bergin and Garvey.

Leacock, Eleanor. 1983. "Ethnohistorical Investigation of Egalitarian Politics in Eastern North America." In *Development of Political Organization in Native North America*, edited by E. Tooker, pp. 17–31. New York: American Ethnological Society.

LeBlanc, Steven A. 1982. "Temporal Change in Mogollon Ceramics." In *Southwestern Ceramics: A Comparative View*, edited by Albert H. Schroeder, pp. 107–128.

Arizona Archaeologist 15, School of American Research Advanced Seminar. Phoenix: Arizona Archaeological Society.
———. 1998. "Settlement Consequences of Warfare During the Late Pueblo III and Pueblo IV Period." In *Migration and Reorganization: The Pueblo IV Period in the American Southwest*, edited by Katherine A. Spielmann, pp. 115–136. Anthropological Research Papers No. 51. Tempe: Arizona State University.
———. 1999. *Prehistoric Warfare in the American Southwest*. Salt Lake City: University of Utah Press.
Lee, Willis T. 1909. *The Manzano Group of the Rio Grande Valley, New Mexico*. U.S. Geological Survey, Bulletin 389. Washington, DC: Government Printing Office.
Lefebvre, Henri. 1991. *The Production of Space*. Translated by Donald Nicholson-Smith. Oxford: Blackwell Publishers.
Lehmer, Donald J. 1948. *The Jornada Branch of the Mogollon*. University of Arizona Bulletin 19(2). Tucson: University of Arizona.
Lekson, Stephen H. 1988. "The Idea of the Kiva in Anasazi Archaeology." *Kiva* 53:213–234.
———. 1999. "Was Casas a Pueblo?" In *The Casas Grandes World*, edited by Curtis F. Schaafsma and Carroll L. Riley, pp. 84–92. Salt Lake City: University of Utah Press.
———. 2002. "War in the Southwest, War in the World." *American Antiquity* 67:607–624.
———. 2006a. "Chaco Matters: An Introduction." In *The Archaeology of Chaco Canyon: An Eleventh-Century Pueblo Regional Center*, edited by Stephen H. Lekson, pp. 3–44. Santa Fe: School of American Research Press.
———. 2006b. "Lords of the Great House: Pueblo Bonito as a Palace." In *Palaces and Power in the Americas: From Peru to the Northwest Coast*, edited by Jessica Joyce Christie and Patricia Joan Sarro, pp. 99–114. Austin: University of Texas Press.
———. 2008. "The Idea of the Kiva in Anasazi Archaeology." *Kiva* 74(2): 203–225.
———. 2009. *A History of the Ancient Southwest*. Santa Fe: School for Advanced Research Press.
Leone, Mark P. 1987. "Rule by Ostentation: The Relationship Between Space and Sight in Eighteenth-Century Landscape Architecture in the Chesapeake Region of Maryland." In *Method and Theory for Activity Area Research*, edited by Susan Kent, pp. 604–633. New York: Columbia University Press.
Liebmann, Matthew, T. J. Ferguson, and Robert W. Preucel. 2005. "Pueblo Settlement, Architecture, and Social Change in the Pueblo Revolt Era, A.D. 1680–1696." *Journal of Field Archaeology* 30:45–60.
Liebmann, Matthew and Robert W. Preucel. 2007. "The Archaeology of the Pueblo Revolt and the Formation of the Modern Pueblo World." *Kiva* 73(2):195–217.
Lightfoot, Kent G. and Gary M. Feinman. 1982. "Social Differentiation and Leadership Development in Early Pithouse Villages in the Mogollon Region of the American Southwest." *American Antiquity* 47:64–86.
Lightfoot, Kent G. and Antoinette Martinez. 1995. "Frontiers and Boundaries in Archaeological Perspective." *Annual Review of Anthropology* 24:471–492.
Lightfoot, Kent G. and Steadman Upham. 1989. "Complex Societies in the Prehistoric Southwest: A Consideration of the Controversy." In *The Sociopolitical Structure of*

Prehistoric Southwestern Societies, edited by Stedman Upham, Kent G. Lightfoot, and Roberta A. Jewett, pp. 3–32. Boulder: Westview Press.

Lightfoot, Ricky R. 1994. *The Duckfoot Site: Archaeology of the House and Household*. Occasional Paper No. 4. Cortez: Crow Canyon Archaeological Center.

Lightfoot, Ricky R. and M. C. Etzkorn, eds. 1993. *The Duckfoot Site, Volume 1: Descriptive Archaeology*. Occasional Paper No. 3. Cortez: Crow Canyon Archaeological Center.

Link, Eugene P. and Beulah M. Link. 1999. *The Tale of Three Cities: Gran Quivira in the Southwest New Mexico, 1100 B.C. to A.D. 1963*. New York: Vantage Press.

Lipe, William D. 1989. "Social Scale of Mesa Verde Anasazi Kivas." In *The Architecture of Social Integration in Prehistoric Pueblos*, edited by William D. Lipe and Michelle Hegmon. Occasional Paper No. 1, pp. 53–71. Cortez: Crow Canyon Archaeological Center.

———. 1994. "Comments on Population Aggregation and Community Organization." In *The Ancient Southwestern Community*, edited by W. H. Wills and Robert D. Leonard, pp. 141–146. Albuquerque: University of New Mexico Press.

Lipe, William D. and Michelle Hegmon, eds. 1989. *The Architecture of Social Integration in Prehistoric Pueblos*. Occasional Paper No. 1. Cortez: Crow Canyon Archaeological Center.

Lovell, Emily K. 1963. "The Lost Mission." *New Mexico Magazine* 41:17–19.

Low, Setha M. 1995. "Indigenous Architecture and the Spanish American Plaza in Mesoamerica and the Caribbean." *American Anthropologist* 97:748–762.

———. 2000. *On the Plaza: The Politics of Public Space and Culture*. Austin: University of Texas Press.

Lowell, Julia C. 1996. "Moieties in Prehistory: A Case Study from the Pueblo Southwest." *Journal of Field Archaeology* 23:77–90.

———. 2007. "Women and Men in Warfare and Migration: Implications of Gender Imbalance in the Grasshopper Region of Arizona." *American Antiquity* 72:95–124.

Lupo, Karen D. 2006. "What Explains the Carcass Field Processing and Transport Decisions of Contemporary Hunter-Gatherers? Measures of Economic Anatomy and Zooarchaeological Skeletal Part Representation." *Journal of Archaeological Method and Theory* 13:19–66.

Lyman, R. Lee. 1985. "Bone Frequencies: Differential Transport, In-Situ Destruction and the MGUI." *Journal of Archaeological Science* 12:221–236.

Lyons, Diane. 1996. "The Politics of House Shape: Round vs. Rectilinear Domestic Structures in Déla Compounds, Northern Cameroon." *Antiquity* 70(268):351–367.

Lyons, Thomas Robert. 1969. "A Study of the PaleoIndian and Desert Culture Complexes of the Estancia Valley Area, New Mexico." Ph.D. dissertation. Department of Anthropology, University of New Mexico, Albuquerque.

Lyons, Thomas R. and Ronald J. Switzer. 1975. "Archaeological Excavations at Tillery Springs, Estancia, NM." In *Collected Papers in Honor of Florence Hawley Ellis*, edited by Theodore R. Frisbie, pp. 312–327. Papers of the Archaeological Society of New Mexico No. 2. Albuquerque: Archaeology Society of New Mexico.

Madsen, Mark, Carl Lipo, and Michael Cannon. 1999. "Fitness and Reproductive Tradeoffs in Uncertain Environments: Explaining the Evolution of Cultural Elaboration." *Journal of Anthropological Archaeology* 18:251–281.

Mahoney, Nancy M., Michael A. Adler, and James W. Kendrick. 2000. "The Changing Scale and Configuration of Mesa Verde Communities." *Kiva* 66:91–122.

Marshall, Michael P. and Henry J. Walt. 1984. *Rio Abajo: Prehistory and History of a Rio Grande Province*. Santa Fe: New Mexico Historic Preservation Program.

Matson, R. G. 1991. *The Origins of Southwestern Agriculture*. Tucson: University of Arizona Press.

Mattessich, Paul and Barbara Monsey. 1997. *Community Building: What Makes It Work*. St. Paul: Amherst H. Wilder Foundation.

McGuire, Randall H. and Dean J. Saitta. 1996. "Although They Have Petty Captains, They Obey Them Badly: The Dialectics of Prehispanic Western Pueblo Social Organization." *American Antiquity* 61:197–216.

———. 1998. "Dialectics, Heterarchy, and Western Pueblo Social Organization." *American Antiquity* 63:334–336.

McGuire, Randall H. and Michael B. Schiffer. 1983. "A Theory of Architectural Design." *Journal of Anthropological Archaeology* 2:277–303.

McWilliams, Kenneth R. 1981. "Non-Metric Oral Traits in Skeletons." In *Contributions to Gran Quivira Archaeology, Gran Quivira National Monument, New Mexico*, edited by Alden C. Hayes, pp. 147–149. National Park Service, Publications in Archeology 17. Washington, DC: U.S. Department of the Interior.

Meinzer, Oscar E. 1911. *Geology and Water Resources of Estancia Valley, New Mexico, with Notes on Ground-Water Conditions in Adjacent Parts of Central New Mexico*. U.S. Geological Survey, Water Supply Paper No. 275. Washington, DC.

Mera, H. P. 1931. *Chupadero Black on White*. Laboratory of Anthropology, Technical Series Bulletin No. 1. Santa Fe: Museum of New Mexico.

———. 1933. A *Proposed Revision of the Rio Grande Plaint Sequence*. Laboratory of Anthropology, Technical Series Bulletin No. 5. Santa Fe: Museum of New Mexico.

———. 1935. *Ceramic Clues to the Prehistory of North Central New Mexico*. Laboratory of Anthropology, Technical Series Bulletin No. 8. Santa Fe: Museum of New Mexico.

———. 1940. *Population Changes in the Rio Grande Glaze-Paint Area*. Laboratory of Anthropology, Technical Series Bulletin No. 9. Santa Fe: Museum of New Mexico.

Miller, Myles R. and Tim B. Graves. 2009. *Madera Quemada Pueblo: Archaeological Investigations of a 14th Century Jornada Mogollon Pueblo*. GMI Report No. 679EP and Fort Bliss Garrison Command, Historic and Natural Resources Report No. 03-12. El Paso: Geo-Marine, Inc.

Mills, Barbara J. ed. 2000. *Alternative Leadership Strategies in the Prehispanic Southwest*. Tucson: University of Arizona Press.

———. 2004a. "The Establishment and Defeat of Hierarchy: Inalienable Possessions and the History of Collective Prestige Structures in the Pueblo Southwest." *American Anthropologist* 106:238–251.

———, ed. 2004b. *Identity, Feasting and the Archaeology of the Greater Southwest*. Proceedings of the 2002 Southwest Symposium. Boulder: University Press of Colorado.

———. 2007. "Performing the Feast: Visual Display and Suprahousehold Commensalism in the Puebloan Southwest." *American Antiquity* 72:210–240.

Mimms, Otho Leroy and William John Zaumeyer. 1947. "Growing Dry Beans in the Western States." *Farmers' Bulletin* No. 1996. Washington, DC: U.S. Department of Agriculture.

Mindeleff, Victor. [1891]1989. *A Study of Pueblo Architecture: Tusayan and Cibola.* Bureau of American Ethnology 8th Annual Report. Washington, DC: Smithsonian Institution Press.

Minnis, Paul E. 1985. *Social Adaptation to Food Stress: A Prehistoric Southwestern Example.* Chicago: University of Chicago Press.

Mobley-Tanaka, Jeannette L. 1997. "Gender and Ritual Space During the Pithouse to Pueblo Transition: Subterranean Mealing Rooms in the North American Southwest." *American Antiquity* 62:437–448.

Mohtashemi, M. and L. Mui. 2003. "Evolution of Indirect Reciprocity by Social Information: The Role of Trust and Reputation in Evolution of Altruism." *Journal of Theoretical Biology* 223:523–531.

Montgomery, John and Kathleen Bowman, eds. 1989. *Archaeological Reconnaissance of the Chupadera Arroyo Drainage, Central New Mexico.* Report CD89, Agency for Conservation Archaeology. Portales: Eastern New Mexico University.

Moore, Jerry D. 1996. "The Archaeology of Plazas and the Proxemics of Ritual: Three Andean Traditions." *American Anthropologist* 98:789–802.

Morgan, Lewis Henry. [1881]1965. *Houses and House-Life of the American Aborigines.* Chicago: University of Chicago Press.

Morgan, Michèle E, ed. 2010. *Pecos Pueblo Revisited: The Biological and Social Context.* Cambridge: Harvard University Press.

Morgan, William. 1994. *Ancient Architecture of the Southwest.* Austin: University of Texas Press.

Morrow, Baker H. 2006. "The Berry Gardens of Quarai and the Pocket Terraces of Abó." In *Canyon Gardens: The Ancient Pueblo Landscapes of the American Southwest,* edited by V. B. Price and Baker H. Morrow, pp. 17–32. Albuquerque: University of New Mexico Press.

Mt. Pleasant, Jane. 2006. "The Science Behind the Three Sisters Mound System: An Agronomic Assessment of an Indigenous Agricultural System in the Northeast." In *Histories of Maize: Multidisciplinary Approaches to the Prehistory, Linguistics, Biogeography, Domestication, and Evolution of Maize,* edited by John E. Staller, Robert H. Tykot, and Bruce F. Benz, pp. 529–537. New York: Academic Press.

Mueller, Aida Linda. 1995. "A Study of the Smudged Utility Ware at Gran Quivira and Quarai Pueblos, New Mexico." B.A. thesis. Archaeology Program, Cornell University, Ithaca, NY.

Murdock, George P. 1949. *Social Structure.* New York: Macmillan.

Murphy, Don. 1993a. *Salinas Pueblo Missions: Abó, Quarai and Gran Quivira.* Tucson: Southwest Parks and Monuments Association.

———. 1993b. "Salinas: A View Through Time." In *Salinas: Archaeology, History, and Prehistory,* edited by David Grant Noble, pp. 6–11. Santa Fe: Ancient City Press.

Murrell, Monica. 2006. *An Investigation of Prehistoric Water Management in the Chupadera Arroyo Basin, Central New Mexico.* Prepared for New Mexico State University Water Resources Research Institute, Las Cruces.

Naroll, Raoul. 1962. "Floor Area and Settlement Population." *American Antiquity* 27(4):587–589.

Nash, Donna J. 2009. "Household Archaeology in the Andes." *Journal of Archaeological Research* 17:205–261.

Nelson, Ben A. 1995. "Complexity, Hierarchy, and Scale: A Controlled Comparison Between Chaco Canyon, New Mexico, and La Quemada, Zacatecas." *American Antiquity* 60:597–618.

Nelson, Ben A., Stephen H. Lekson, Ivan Sprajc, and Kenneth E. Sassaman. 2011. "Shaping Space: Built Space, Landscape, and Cosmology in Four Regions." *Santa Fe Institute Working Paper* 2011-02-005.

Nelson, Margaret C. and Michelle Hegmon. 2001. "Abandonment Is Not as It Seems: An Approach to the Relationship Between Site and Regional Abandonment." *American Antiquity* 66:213–235.

Nelson, Margaret C., Michelle Hegmon, Stephanie Kulow, and Karen Gust Schollmeyer. 2006. "Archaeological and Ecological Perspectives on Reorganization: A Case Study from the Mimbres Region of the U.S. Southwest." *American Antiquity* 71:403–432.

Nelson, Margaret C. and Gregson Schachner. 2002. "Understanding Abandonments in the North American Southwest." *Journal of Archaeological Research* 10(2):167–206.

Netting, Robert M. 1990. "Population, Permanent Agriculture, and Polities: Unpacking the Evolutionary Portmanteau." In *The Evolution of Political Systems: Sociopolitics in Small-Scale Societies*, edited by Steadman Upham, pp. 21–61. Cambridge: Cambridge University Press.

Netting, Robert M., Richard R. Wilk, and Eric J. Arnud, eds. 1984. *Households: Comparative and Historical Studies of the Domestic Group*. Berkeley: University of California Press.

New Mexico Interstate Stream Commission and the New Mexico State Engineer Office. 1974. *County Profile, Torrance County, New Mexico. Water Resources Assessment for Planning Purposes*. Santa Fe: New Mexico Interstate Stream Commission and the New Mexico State Engineer Office.

New York Times. 1988. "Fredrico Sisneros." Obituary accessed online at http://www.nytimes.com/1988/03/15/obituaries/federico-sisneros-93-longtime-park-ranger.html.

Newman, O. 1972. *Defensible Space*. New York: Macmillan.

Nichols, Deborah L. and Patricia L. Crown, eds. 2008. *Social Violence in the Prehispanic American Southwest*. Tucson: University of Arizona Press.

Nielsen, Axel. 1995. "Architectural Performance and the Reproduction of Social Power." In *Expanding Archaeology*, edited by James M. Skibo, William H. Walker, and Axel E. Nielsen, pp. 47–66. Salt Lake City: University of Utah Press.

Noble, David Grant. 1993. *Salinas: Archaeology, History, and Prehistory*. Santa Fe: Ancient City Press.

Nowak, M. A. 2006. "Five Rules of the Evolution of Cooperation." *Science* 314: 1560–1563.

Nowak, M. A. and K. Sigmund. 2005. "Evolution of Indirect Reciprocity." *Nature* 437: 1291–1298.

Olsen, John W. 1990. *Vertebrate Faunal Remains from Grasshopper Pueblo, Arizona*. Anthropological Papers 83. Ann Arbor: University of Michigan Museum of Anthropology.

Ortman, Scott G. 1998. "Corn Grinding and Community Organization in the Pueblo Southwest, A.D. 1150–1550." In *Migration and Reorganization: The Pueblo IV Period in the American Southwest*, edited by Katherine A. Spielmann, pp. 165–192. Anthropological Research Papers No. 51. Tempe: Arizona State University.

Ortman, Scott G., Donna M. Glowacki, Melissa J. Churchill, and Kristin A. Kuckelman. 2000. "Pattern and Variation in Northern San Juan Village Histories." *Kiva* 66:123–146.

Otterbein, Keith F. 2004. *How War Began*. College Station: Texas A&M University Press.

———. 2009. *The Anthropology of War*. Boulder: Westview Press.

Palmer, Craig T., B. Eric Fredrickson, and Christopher F. Tilley. 1997. "Categories and Gatherings: Group Selection and the Mythology of Cultural Anthropology." *Evolution and Human Behavior* 18:291–308.

Parker, Bradley J. 2006. "Toward an Understanding of Borderland Processes." *American Antiquity* 71:77–100.

Parsons, Francis B. 1975. *Early 17th Century Missions of the Southwest*. Tucson: Dale Stuart King, Publisher.

Peckham, Stewart. 1976. "Taylor Draw: A Mogollon-Anasazi Hybrid?" In *Collected Papers in Honor of Marjorie Ferguson Lambert*, edited by Albert H. Schroeder, pp. 37–71. Papers of the Archaeological Society of New Mexico No. 3. Albuquerque: Archaeological Society of New Mexico.

Peeples, Matthew. 2014. "Population History of the Zuni Region Across the Protohistoric Transition: Migration, Gene Flow, and Social Transformation." In *Building Transnational Archaeologies*, edited by Randall H. McGuire and Elisa Villalpando. Arizona State Museum Archaeology Series.

Peregrine, Peter N. 2001. "Matrilocality, Corporate Strategy, and the Organization of Production in the Chacoan World." *American Antiquity* 66:36–46.

Phillips, David A., Jr., and Lynne Sebastian. 2004. "Large-Scale Feasting and Politics: An Essay on Power in Precontact Southwestern Societies." In *Identity, Feasting, and the Archaeology of the Greater Southwest*. Proceedings of the 2002 Southwest Symposium, edited by Barbara J. Mills, pp. 233–260. Boulder: University Press of Colorado.

Plog, Stephen. 1997. *Ancient Peoples of the American Southwest*. New York: Thames and Hudson.

Plog, Stephen and Julie Solometo. 1997. "The Never-Changing and the Ever-Changing: The Evolution of Western Pueblo Ritual." *Cambridge Archaeological Journal* 7(2):161–182.

Potter, James. 2000. "Pots, Parties, and Politics: Communal Feasting in the American Southwest." *American Antiquity* 65:471–492.

Potter, James M. and Jason Chuipka. 2007. "Early Pueblo Communities and Cultural Diversity in the Durango Area: Preliminary Results from the Animas–La Plata Project." *Kiva* 72:407–430.

Price, T. Douglas, Clark M. Johnson, Joseph A. Ezzo, Jonathan Ericson, and James H. Burton. 2000. "Residential Mobility in the Prehistoric Southwest United States: A Preliminary Study Using Strontium Isotope Analysis." *Journal of Archaeological Science* 21:315–330.

Prince, L. Bradford. 1915. *Spanish Mission Churches of New Mexico*. Cedar Rapids: The Torch Press. Accessed online through *Books of the Southwest: The University of Arizona Library*. June 2009. http://southwest.library.arizona.edu/spmc/index.html.

Prudden, Mitchell. 1914. "The Circular Kivas of Small Ruins in the San Juan Watershed." *American Anthropologist* 16:33–58.

Psarra, Sophia. 2009. *Architecture and Narrative: The Formation of Space and Cultural Meaning*. New York: Routledge.

Puseman, Kathryn and Alison E. Rautman. 2012. "Macrofloral Remains and Radiocarbon Dates from Frank's Pueblo, LA 9032, New Mexico." Poster presented at the Society for American Archaeology 77th Annual Meeting, Memphis, Tennessee.

Rapoport, Amos. 1969 "The Pueblo and the Hogan: A Cross-Cultural Comparison of Two Responses to an Environment." In *Shelter and Society*, edited by P. Oliver, pp. 66–70. London: Barrie and Rockliff.

———. 1982. *The Meaning of the Built Environment: A Nonverbal Communication Approach*. Beverly Hills: Sage Publications.

———. 1988. "Levels of Meaning in the Built Environment." In *Cross-Cultural Perspectives in Non Verbal Communication*, edited by Fernando Poyatos, pp. 317–336. Toronto: C. J. Hogrefe.

Rautman, Alison E. 1990. "The Environmental Context of Decision-Making: Coping Strategies Among Prehistoric Cultivators in Central New Mexico." Ph.D. dissertation. Department of Anthropology, University of Michigan, University Microfilms, Ann Arbor.

———. 1991. "Excavations at LA 38448, the Kite Site, a Pithouse Village in the Salinas Region of Central New Mexico." In *Jornada Mogollon Archaeology: Collected Papers from the Fifth and Sixth Jornada Mogollon Conferences*, edited by Meliha S. Duran and Patrick H. Beckett. Las Cruces: COAS Publishing and Research.

———. 1992. *A Preliminary Report of the 1992 Excavations at LA 2091, Pueblo de la Mesa, Cibola National Forest, New Mexico*. Manuscript on file, USDA Forest Service, Cibola National Forest, Albuquerque.

———. 1993a. *A Preliminary Report of the 1993 Field Season at LA 2091, Pueblo de la Mesa, Cibola National Forest, New Mexico*. Manuscript on file, USDA Forest Service, Cibola National Forest, Albuquerque.

———. 1993b. "Resource Variability, Risk, and the Structure of Social Networks: An Example from the Prehistoric Southwest." *American Antiquity* 58(3):403–424.

———. 1994. "Regional Climate Records and Local Experience: Drought and the Decline of Dryfarming in Central New Mexico." *Culture and Agriculture* 49:12–15.

———. 1995a. *The 1994 Field Season at LA 199, Kite Pueblo, Torrance County, New Mexico*. Manuscript on file, Museum of New Mexico Laboratory of Anthropology (Santa Fe) and with Salinas Pueblo Missions National Monument (Mountainair).

———. 1995b. *Report on the Excavation of Human Remains from the 1994 Field Season at LA 199, Kite Pueblo, Torrance County, New Mexico*. Manuscript on file, State of New Mexico Office of Cultural Affairs, Historic Preservation Division, Santa Fe.

———. 1995c. *Final Report on the Excavation of Pueblo de la Mesa, Cibola National Forest, Torrance County, New Mexico*. Manuscript on file. USDA Forest Service, Cibola National Forest, Albuquerque.

———. 1996. "Risk, Reciprocity, and the Structure of Social Networks." In *Evolving Complexity and Environmental Risk in the Prehistoric Southwest*, edited by J. Tainter and B. B. Tainter. Santa Fe Institute Studies in the Science of Complexity, Vol. 24, pp. 197–222. Reading: Addison-Wesley.

———. 1997. "The Pithouse to Pueblo Transition in the American Southwest: Implications for Gender Roles." In *Women in Prehistory: North America and Mesoamerica*, edited by Cheryl Claassen and Rosemary A. Joyce, pp. 100–118. Philadelphia: University of Pennsylvania Press.

———. 2000. "Population Aggregation, Community Organization, and Plaza Oriented Pueblos in the American Southwest." *Journal of Field Archaeology* 27:271–283.

———. 2008a. "Two Centuries of Heritage Tourism at Gran Quivira National Monument." Paper presented at the Society for Historic Archaeology, Albuquerque, New Mexico.

———. 2008b. *Preliminary Report of the 2007 Excavations at Frank's Ruin (LA 9032), New Mexico*. Manuscript on file, State of New Mexico Office of Cultural Affairs, Historic Preservation Division, Santa Fe.

———. 2011. "Village and Community in New Mexico." Paper presented at the 76th Annual Meeting of the Society for American Archaeology, Sacramento, California.

———. 2013a. "Social Integration and the Built Environment of Aggregated Communities in the North American Puebloan Southwest." In *From Prehistoric Villages to Cities: Settlement Aggregation and Community Transformation*, edited by Jennifer Birch, pp. 109–133. New York: Routledge.

———. 2013b. "Household, Community, and Circular Pueblos in the American Southwest: A Case Study from the Salinas Region of Central New Mexico." Paper presented at the Society for American Archaeology, Honolulu, Hawaii.

———. 2013c. *Final Report of Excavations at LA 9032, Frank's Pueblo*. Manuscript on file, State of New Mexico Department of Cultural Affairs, Historic Preservation Division, Santa Fe.

Rautman, Alison E. and Matthew A. Chamberlin. 2008. "When Is a Defensive Site? Coping with Conflict in Central New Mexico." Poster presented at the 73rd Annual Meeting of the Society for American Archaeology, Vancouver, British Columbia.

———. 2010. "Interpreting Burning in the Salinas Pueblo Province, A.D. 1275–1540." Paper presented at the 75th Annual Meeting of the Society for American Archaeology, St. Louis, Missouri.

Rautman, Alison E. and Anne M. Colyer. 1997. "Economic Organization During the Pithouse-to-Pueblo Transition: Evidence from Stable Isotope Analysis of Human Bone." Paper presented at the 62nd Annual Meeting of the Society for American Archaeology, Nashville, Tennessee.

Rautman, Alison E. and Marjorie Heyman. 2011. *Final Report on the Excavation of Kite Pueblo, LA 199, Torrance County, NM*. Report on file, Museum of New Mexico, Santa Fe.

Reed, Erik K. 1940. "History of Quarai." In *Special Report on Quarai State Monument*. May 1939, downloaded December 2007. http://www.nps.gov/history/history/online_books/sapu/quarai-history.htm.

———. 1956. "Types of Village-Plan Layouts in the Southwest." In *Prehistoric Settlement Patterns in the New World*, edited by Gordon R. Willey, pp. 11–17. Viking Fund Publications in Anthropology 23. New York: Viking Fund.

———. 1981. "Human Skeletal Material from the Gran Quivira District." In *Contributions to Gran Quivira Archaeology*, edited by Alden C. Hayes, pp. 75–118. National Park Service, Publications in Archeology 17. Washington, DC: U.S. Department of the Interior.

Reed, Paul F., ed. 2000. *Foundations of Anasazi Culture: The Basketmaker-Pueblo Transition*. Salt Lake City: University of Utah Press.

Reese-Taylor, K. and R. Koontz. 2001. "The Cultural Poetics of Power and Space in Ancient Mesoamerica." In *Landscape and Power in Ancient Mesoamerica*, edited by R. Koontz, K. Reese-Taylor, and A. Headrick, pp. 1–28. Boulder: Westview Press.

Riggs, Charles R. 2001. *The Architecture of Grasshopper Pueblo*. Salt Lake City: University of Utah Press.

———. 2007. "Architecture and Identity at Grasshopper Pueblo, Arizona." *Journal of Archaeological Research* 63(4):498–513.

Robb, John. 2007. *The Early Mediterranean Village: Agency, Material Culture, and Social Change in Neolithic Italy*. Cambridge: Cambridge University Press.

Robin, Cynthia and Nan A. Rothschild. 2002. "Archaeological Ethnographies: Social Dynamics of Outdoor Space." *Journal of Social Archaeology* 2(2):159–172.

Rocek, Thomas R. 1995. "Sedentarization and Agricultural Dependence: Perspectives from the Pithouse-to-Pueblo Transition in the American Southwest." *American Antiquity* 60:218–239.

Rocek, Thomas R. and Alison E. Rautman. 2007. "No Peripheral Vision: A View of Regional Interactions from South-Central New Mexico." In *Hinterlands and Regional Dynamics in the Ancient Southwest*, edited by Alan P. Sullivan III and James M. Bayman, pp. 125–138. Tucson: University of Arizona Press.

———. 2012. "First-Millennium Pithouse Community Diversity in Southeastern New Mexico." In *Southwestern Pithouse Communities, AD 200–900*, edited by Lisa C. Young and Sarah A. Herr, pp. 110–122. Tucson: University of Arizona Press.

Rogers, Richard. 1980. "Chemistry of Pottery Smudging." *Pottery Southwest* 2–4.

Rogoff, Barbara. 2003. *The Cultural Nature of Human Development*. Oxford: Oxford University Press.

Roosa, William. B. 1956a. "Preliminary Report on the Lucy Site." *El Palacio* 63(2):36–49.

———. 1956b. "The Lucy Site in Central New Mexico." *American Antiquity* 21:310.

Roscoe, Paul. 2009. "Social Signaling and the Organization of Small-Scale Society: The Case of Contact-Era New Guinea." *Journal of Archaeological Method and Theory* 16:69–116.

Roth, Barbara J., ed. 2010. *Engendering Households in the Prehistoric Southwest*. Tucson: University of Arizona Press.

Roth, Barbara J. and Robert J. Stokes, eds. 2007. *Exploring Variability in Mogollon Pithouses*. Anthropological Research Papers No. 58. Tempe: Arizona State University.

Ruscavage-Barz, Samantha M. and Elizabeth A. Bagwell. 2006. "Gathering Spaces and Bounded Places: The Religious Significance of Plaza-Oriented Communities in the Northern Rio Grande, New Mexico." In *Religion in the Prehispanic Southwest*, edited by Christine S. Van Pool, Todd L. Van Pool, and David A. Phillips, Jr., pp. 81–102. New York: Altamira Press.

Russo, Michael. 2004. "Measuring Shell Rings for Social Inequality." In *Signs of Power: The Rise of Cultural Complexity in the Southeast*, edited by Jon L. Gibson and Philip J. Carr, pp. 26–70. Tuscaloosa: University of Alabama Press.

Sahlins, Marshall D. 1972. *Stone Age Economics*. New York: Aldine Publishing Company.

Sanchez, Joseph P. 1993. "Bernardo Gruber and the New Mexico Inquisition." In *Salinas: Archaeology, History, and Prehistory*, edited by David Grant Noble, pp. 26–31. Santa Fe: Ancient City Press.

Scarborough, Vernon L. 1988. "A Water Storage Adaptation in the American Southwest." *Journal of Anthropological Research* 44:21–40.

———. 1989. "Site Structure of a Village of the Late Pithouse–Early Pueblo Period in New Mexico." *Journal of Field Archaeology* 16:405–425.

Schaafsma, Polly. 1993. "Tompiro Rock Art." In *Salinas: Archaeology, History, and Prehistory*, edited by David Grant Noble, pp. 16–19. Santa Fe: Ancient City Press.

Schachner, Gregson. 2001. "Ritual Control and Transformation in Middle-Range Societies: An Example from the American Southwest." *Journal of Anthropological Archaeology* 20: 168–194.

———. 2010. "Corporate Group Formation and Differentiation in Early Puebloan Villages of the American Southwest." *American Antiquity* 75:473–496.

———. 2012. *Population Circulation and the Transformation of Ancient Zuni Communities*. Tucson: University of Arizona Press.

Schaepe, David M. 2006. "Rock Fortification: Archaeological Insights into Precontact Warfare and Sociopolitical Organization Among the Sto-Lo of the Lower Fraser River Canyon, B.C." *American Antiquity* 71:671–706.

Schiffer, M. B. and R. H. McGuire. 1992. "A Theory of Architectural Design." In *Technological Perspectives on Behavioral Change*, edited by Michael Schiffer, pp. 22–43. Tucson: University of Arizona Press.

Schlanger, Sarah. 1992. "Recognizing Persistent Places in Anasazi Settlement Systems." In *Space, Time, and Archaeological Landscapes*, edited by R. Koontz, K. Reese-Taylor, and A. Headrick, pp. 29–53. Boulder: Westview Press.

Schroeder, A. H. 1979. "Pueblos Abandoned in Historic Times." In *Handbook of North American Indians, Volume 9: Southwest*, edited by Alfonso Ortiz and William C. Sturtevant, pp. 236–254. Washington, DC: Smithsonian Institution Press.

Senter, Donovan. 1934. "The Work on the Old Quarai Mission." *El Palacio* 37(21-22-23):169–174.

Seymour, Deni J. 2009. "Distinctive Places, Suitable Spaces: Conceptualizing Mobile Group Occupation Duration and Landscape Use." *International Journal of Historical Archaeology* 13:255–281.

Shafer, Harry. 1995. "Architecture and Symbolism in Transitional Pueblo Development in the Mimbres Valley, Southwest New Mexico." *Journal of Field Archaeology* 22(1):23–47.

Shanks, Michael and Randall H. McGuire. 1996. "The Craft of Archaeology." *American Antiquity* 61:75–88.

Shapiro, Jason S. 2005. *A Space Syntax Analysis of Arroyo Hondo Pueblo, New Mexico: Community Formation in the Northern Rio Grande*. Santa Fe: School of American Research Press.

Shepard, Anna O. 1936. "The Technology of Pecos Pottery." In *The Pottery of Pecos, Volume II,* edited by A. V. Kidder and Anna O. Shepard, pp. 389–587. New Haven: Yale University Press.

———. 1942. *Grande Glaze Paint Ware: A Study Illustrating the Place of Ceramic Technological Analysis in Archaeological Research.* Contributions to American Anthropology and History 39, Publication 528. Washington, DC: Carnegie Institute of Washington.

———. 1954. *Ceramics for the Archaeologist.* Publication 609. Washington, DC: Carnegie Institute of Washington.

Sholes, France V. 1940. "Troublous Times in New Mexico." *New Mexico Historical Review* 15:249–268.

Smith, Michael E. 1983. "Pueblo Councils: An Example of Stratified Egalitarianism." In *The Development of Political Organization in Native North America,* edited by Elizabeth Tooker, pp. 32–44. Washington, DC: American Ethnological Society.

Smith, Watson. 1990. *When Is a Kiva? And Other Questions About Southwestern Archaeology.* Edited by Raymond H. Thompson. Tucson: University of Arizona Press.

Snead, James. 2008. *Ancestral Landscapes of the Puebloan World.* Tucson: University of Arizona Press.

Snead, James E. and Robert W. Preucel. 1999. "The Ideology of Settlement: Ancestral Keres Lanscapes in the Northern Rio Grande." In *Archaeology of Landscapes: Contemporary Perspectives,* edited by Wendy Ashmore and A. Bernard Knapp, pp. 169–197. Malden: Blackwell Publishers.

Snow, David H. 1973. "Some Economic Considerations of Historic Rio Grande Pueblo Pottery." In *The Changing Ways of Southwestern Indians, A Historic Perspective,* edited by A. H. Schroeder, pp. 55–72. Glorieta: The Rio Grande Press.

———. 2007. "Recombining D 'n A: Rio Grand Glaze-Ware Sequence Revisited." Paper presented at the Pecos Conference, Pecos National Historic Park, New Mexico.

Solometo, Julie P. 2004. "The Conduct and Consequences of War: Dimension of Conflict in East Central Arizona." Ph.D. dissertation. Department of Anthropology, University of Michigan, Ann Arbor.

———. 2006. "The Dimensions of War: Conflict and Culture Change in Central Arizona." In *The Archaeologies of Warfare: Prehistories of Raiding and Conquest,* edited by Elizabeth N. Arkush and Mark W. Allen, pp. 23–65. Gainesville: University Press of Florida.

Solometo, Julie P., Alison E. Rautman, and Matthew A. Chamberlin. 2011. "Assumptions, Expectations and the (Apparent) Reality of 12th and 13th Century Village Life in Central New Mexico." Paper presented at the 76th Annual Meeting of the Society for American Archaeology, Sacramento, California.

Sommer, Robert. 1969. *Personal Space: The Behavioral Basis for Design.* New York: Prentice-Hall.

Speth, John D. and Alison E. Rautman. 2004. "Bison Hunting at the Henderson Site." In *Life on the Periphery: Economic Change in Late Prehistoric Southeastern New Mexico,* edited by John D. Speth, pp. 107–206. Memoirs of the Museum of Anthropology. Ann Arbor: University of Michigan.

Spicer, Edward H. [1961]1981. *Cycles of Conquest.* Tucson: University of Arizona Press.

Spielmann, Katherine A. 1991. "Coercion or Cooperation? Plains-Pueblo Interaction in the Protohistoric Period." In *Farmers, Hunters, and Colonists: Interaction Between the Southwest and the Southern Plains*, edited by Katherine A. Spielmann, pp. 36–50. Tucson: University of Arizona Press.

———. 1992. *Subsistence and Exchange at Gran Quivira Pueblo. Part I and II*. Manuscript submitted to the Southwestern Regional Office of the National Park Service, Santa Fe, New Mexico.

———. 1993. *The Evolution of Craft Specialization in Tribal Societies: Preliminary Report of the 1992 Excavation Season at Quarai Pueblo, New Mexico*. Manuscript submitted to the Southwestern Regional Office of the National Park Service, Santa Fe, New Mexico.

———. 1994a. *The Evolution of Craft Specialization in Tribal Societies: Preliminary Report for the 1993 Excavation Season at Quarai Pueblo, New Mexico*. Manuscript submitted to the Southwestern Regional Office of the National Park Service, Santa Fe, New Mexico.

———. 1994b. "Clustered Confederacies: Sociopolitical Organization in the Protohistoric Rio Grande." In *The Ancient Southwestern Community: Models and Methods of the Study of Prehistoric Social Organization*, edited by W. H. Wills and R. D. Leonard, pp. 45–54. Albuquerque: University of New Mexico Press.

———. 1996. "Impressions of Pueblo III Settlement Trends Among the Rio Abajo and Eastern Border Pueblos." In *The Prehistoric Pueblo World, A.D. 1150–1350*, edited by Michael A. Adler, pp. 177–187. Tucson: University of Arizona Press.

———. 1998a. *Economy and Society at Pueblo Colorado, New Mexico: Report of the 1989 Excavations*. Report prepared for the USDA Forest Service, Cibola Office, Albuquerque, New Mexico.

———. 1998b. "Ritual Influences on the Development of Rio Grande Glaze A Ceramics." In *Migration and Reorganization: The Pueblo IV Period in the American Southwest*, edited by Katherine A. Spielmann, pp. 253–264. Anthropological Research Papers No. 51. Tempe: Arizona State University.

———, ed. 1998c. *Migration and Reorganization: The Pueblo IV Period in the American Southwest*. Anthropological Research Papers No. 51. Tempe: Arizona State University.

———. 2002. "Feasting, Craft Specialization, and the Ritual Mode of Production in Small-Scale Societies." *American Anthropologist* 104:195–207.

———. 2004a. "Communal Feasting, Ceramics, and Exchange." In *Identity, Feasting, and the Archaeology of the Greater Southwest*. Proceedings of the 2002 Southwest Symposium, edited by Barbara J. Mills, pp. 210–232. Boulder: University Press of Colorado.

———. 2004b. "Clusters Revisited." In *The Protohistoric Pueblo World, A.D. 1275–1600*, edited by Charles Adams and Andrew Duff, pp. 137–145. Tucson: University of Arizona Press.

———. 2011. Salinas Pueblo Nucleation. *Archaeology Southwest* 25:6–7.

Spielmann, Katherine A. and David C. Eshbaugh. 1989. *Summary Report: Salinas Archaeological Survey, May–June 1988*. Manuscript on file at Department of Anthropology, Arizona State University, and at Salinas Pueblo Missions National Monument, Mountainair, New Mexico.

Spielmann, Katherine A., Tiffany Clark, D. Hawkey, Kathrine Rainey, and Susan K. Fish. 2008. "'. . . being weary, they had rebelled': Pueblo Subsistence and Labor under Spanish Colonialism." *Journal of Anthropological Archaeology* 28(1): 102–125.

Spielmann, Katherine A., Jeannette L. Mobley-Tanaka, and James M. Potter. 2006. "Style and Resistance in the Seventeenth-Century Salinas Province." *American Antiquity* 71:621–648.

Spier, Leslie. 1917. *An Outline for a Chronology of Zuni Ruins*. New York: Columbia University Press.

Squires, Ida Belle. 1923. *Field Work Summer of 1923*. Manuscript on file, Salinas Pueblo Missions National Monument, Mountainair, New Mexico.

Staller, John E., Robert H. Tykot, and Bruce F. Benz, eds. 2006. *Histories of Maize: Multidisciplinary Approaches to the Prehistory, Linguistics, Biogeography, Domestication, and Evolution of Maize*. New York: Academic Press.

Stanley, Andy and Bill Willits. 2004. *Creating Community: Five Keys to Building a Small Group Culture*. North Point Resources. Colorado Springs: Multnomah Books.

Stark, Miriam, ed. 1998. *The Archaeology of Social Boundaries*. Washington, DC: Smithsonian Institution Press.

Steadman, Sharon R. 1996. "Recent Research in the Archaeology of Architecture: Beyond the Foundations." *Journal of Archaeological Research* 4:51–93.

Stevenson, Matilda Coxe. 1904. *The Zuni Indians: Their Mythology, Esoteric Societies, and Fraternities*. Twenty-Third Annual Report of the Bureau of American Ethnology, 1901–1902. Washington, DC: U.S. Government Printing Office.

Steward, Julian H. 1937. "Ecological Aspects of Southwestern Society." *Anthropos* 32:87–104.

Stodder, Ann. 2012. "Quantifying Morbidity and Quality of Life in the Prehispanic Southwestern Villages." Paper presented at the 13th Southwest Symposium, Albuquerque, New Mexico.

Stokes, Robert J. and Barbara J. Roth. 1999. "Mobility, Sedentism and Settlement Patterns in Transition: the Late Pithouse Period in the Sapillo Valley, New Mexico." *Journal of Field Archaeology* 26:423–434.

Stone, Glenn Davis. 1991. "Agricultural Territories in a Dispersed Settlement System." *Current Anthropology* 32:343–353.

Stone, Glenn Davis and Christian E. Downum. 1999. "Non-Boserupian Ecology and Agricultural Risk: Ethnic Politics and Land Control in the Arid Southwest." *American Anthropologist* 101(1):113–128.

Stone, Tammy. 2000. "Prehistoric Community Integration in the Point of Pines Region of Arizona." *Journal of Field Archaeology* 27(2):197–208.

———. 2003. "Social Identity and Ethnic Interaction in the Western Pueblos of the American Southwest." *Journal of Archaeological Method and Theory* 10(1):31–67.

Stuart, David E. 2000. *Anasazi America: Seventeen Centuries on the Road from Center Place*. Albuquerque: University of New Mexico Press.

Stuart, David E. and Robin Y. Farwell. 1983. "Out of Phase: Late Pithouse Occupations in the Highlands of New Mexico." In *High Altitude Adaptations in the Southwest*, edited by Joseph C. Winter, pp. 115–159. Cultural Resource Management Report 2. Albuquerque: U.S. Department of Agriculture.

Stuart, David E. and Rory P. Gauthier. 1981. *Prehistoric New Mexico: Background for Survey*. Santa Fe: Historic Preservation Bureau.

Stubbs, Stanley A. 1959. "New Old Churches Found at Quarai and Tabira (Pueblo Blanco)." *El Palacio* 66:162–169.

Stuiver, M. and B. Becker. 1986. "High-Precision Decadal Calibration of the Radiocarbon Time Scale, A.D. 1950–2500 B.C." *Radiocarbon* 28:863–910.

Szuter, Christine R. and Frank Bayham. 1989. "Sedentism and Animal Procurement Among Desert Horticulturalists of the North American Southwest." In *Farmers as Hunters: The Implications of Sedentism*, edited by Susan Kent, pp. 80–95. Cambridge: Cambridge University Press.

Tainter, Joseph A. 1979. "The Mountainair Lithic Scatters: Settlement Patterns and Significance Evaluation of Low Density Surface Sites." *Journal of Field Archaeology* 6:463–469.

Tainter, Joseph A. and Frances Levine. 1987. *Cultural Resources Overview: Central New Mexico*. Albuquerque: U.S. Department of Agriculture Forest Service and Bureau of Land Management.

Testart, Alain. 1982. "The Significance of Food Storage Among Hunter-Gatherers: Residence Patterns, Population Densities, and Social Inequalities." *Current Anthropology* 23(5):523–537.

Titiev, Mischa. 1944. *Old Oraibi, A Study of Hopi Indians of Third Mesa*. Papers of the Peabody Museum of American Archaeology and Ethnology, Vol. 22, No. 1. Cambridge: Harvard University.

Toulouse, Jr., Joseph H. 1945. "Early Water Systems at Gran Quivira National Monument." *American Antiquity* 10:362–372.

———. 1949. *The Mission of San Gregorio de Abo*. Monograph 13. Santa Fe: School of American Research.

Toulouse, Jr., Joseph H. and Robert L. Stephenson. 1960. *Excavations at Pueblo Pardo*. Papers in Anthropology 2. Santa Fe: The Museum of New Mexico.

Trigg, Heather. 1995. "Ethnobotanical Remains." In *Final Report on the Excavation of Pueblo de la Mesa, Cibola National Forest, Torrance County, New Mexico*, compiled by Alison E. Rautman. U.S.D.A. Forest Service, Cibola National Forest, Albuquerque, New Mexico.

Turner, Christy G., II. 1981. "The ASU Study of Gran Quiviran Physical Anthropology." In *Contributions to Gran Quivira Archeology, Gran Quivira National Monument, New Mexico*, edited by Alden C. Hayes, pp. 119–122. National Park Service, Publications in Archeology 17. Washington, DC: U.S. Department of the Interior.

Van Dyke, Ruth. 2008. *The Chaco Experience: Landscape and Ideology at the Center Place*. Santa Fe: School of American Research Press.

Varien, Mark D. 2002. "Persistent Communities and Mobile Households: Population Movement in the Central Mesa Verde Region, A.D. 950 to 1290." In *Seeking the Center Place: Archaeology and Ancient Communities in the Mesa Verde Region*, edited by Mark D. Varien and Richard H. Wilshusen, pp. 163–184. Salt Lake City: University of Utah Press.

Varien, Mark D. and Ricky R. Lightfoot. 1989. "Ritual and Nonritual Activities in Mesa Verde Region Pit Structures." In *The Architecture of Social Integration in*

Prehistoric Pueblos, edited by William D. Lipe and Michelle Hegmon, pp. 73–88. Occasional Paper No. 1. Cortez: Crow Canyon Archaeological Center.

Varien, Mark D. and Scott G. Ortman. 2005. "Accumulations Research in the Southwest United States: Middle-Range Theory for Big-Picture Problems." *World Archaeology* 37:132–155.

Varien, M. D. and J. M. Potter, eds. 2008. *The Social Construction of Communities: Agency, Structure, and Identity in the Prehistoric Southwest*. Lanham: AltaMira Press.

Veblen, Thorstein. 1899. *Theory of the Leisure Class*. New York: The Macmillan Company.

Vivian, Gordon. 1979. *Gran Quivira: Excavations in a 17th Century Jumano Mission*. National Park Service, Archaeological Research Series 8. Washington, DC: U.S. Department of the Interior.

Vlasich, James A. 2005. *Pueblo Indian Agriculture*. Albuquerque: University of New Mexico Press.

Wait, Walter K. and Peter J. Mckenna. 1990. *Quarai Parking Lot Rehabilitation Archeological Testing Program Salinas Pueblo Missions National Monument*. Southwest Cultural Resources Professional Papers No. 27. Manuscript on file. Salinas Pueblo Missions National Monument, Mountainair, New Mexico.

Wallace, Jennifer. 2004. *Digging the Dirt: The Archaeological Imagination*. London: Duckworth.

Walter, Paul A. F. 1931. *The Cities that Died of Fear*. School of American Research. Santa Fe: El Palacio Press.

Ware, John A. 2002. "What Is a Kiva? The Social Organization of Early Pueblo Communities." In *Culture and Environment in the American Southwest: Essays in Honor of Robert C. Euler*, edited by David A. Phillips Jr. and John A. Ware, pp. 79–88. SWCA Anthropological Research Paper No. 8. Tucson: University of Arizona Press.

Ware, John A. and Eric Blinman. 2000. "Cultural Collapse and Reorganization: Origin and Spread of Pueblo Ritual Sodalities." In *The Archaeology of Regional Interaction: Religion, Warfare, and Exchange Across the American Southwest and Beyond*, edited by Michele Hegmon, pp. 381–409. Boulder: University Press of Colorado.

Warren, A. Helene. 1980. "Prehistoric Pottery of Tijeras Canyon." In *Tijeras Canyon: Analysis of the Past*, edited by Linda S. Cordell, pp. 149–168. Albuquerque: University of New Mexico Press.

———. 1981. "A Petrographic Study of the Pottery of Gran Quivira." In *Contributions to Gran Quivira Archaeology, Gran Quivira National Monument, New Mexico*, edited by Alden C. Hayes, pp. 57–74. National Park Service, Publications in Archeology 17. Washington, DC: U.S. Department of the Interior.

Weber, Robert H. 1973. "The Tajo 2 Pit-House Site near Socorro, New Mexico." *Awanyu* 1(1):14–20.

———. 1997. "Geology of Mockingbird Gap Site in Central New Mexico." In *Layers of Time, Papers in Honor of Robert H. Weber*, edited by Meliha S. Duran and David T. Kirkpatrick, pp. 115–122. Albuquerque: The Archaeological Society of New Mexico.

Weber, Robert. H. and George A. Agogino. 1968. "Mockingbird Gap Paleo-Indian Site: Excavations in 1967." Paper presented at the 33rd Annual Meeting of the Society for American Archaeology, Santa Fe, New Mexico.

———. 1997. "Mockingbird Gap PaleoIndian Site: Excavations in 1967." In *Layers of Time, Papers in Honor of Robert H. Weber*, edited by Meliha S. Duran and David T. Kirkpatrick, pp. 123–127. Albuquerque: The Archaeological Society of New Mexico.

Weeks, John M. 1988. "Residential and Local Group Organization in the Maya Lowland of Southwestern Campeche, Mexico." In *Household and Community in the Mesoamerican Past*, edited by Richard R. Wilk and Wendy Ashmore, pp. 73–96. Albuquerque: University of New Mexico Press.

Wesson, Cameron B. 1999. "Chiefly Power and Food Storage in Southeastern North America." *World Archaeology* 31:145–164.

Whalen, Michael E. 1981a. "Cultural-Ecological Aspects of the Pithouse-to-Pueblo Transition in a Portion of the Southwest." *American Antiquity* 46:75–92.

———. 1981b. "An Investigation of Pithouse Village Structure in Western Texas." *Journal of Field Archaeology* 8:303–311.

———. 1990. "Defining Buried Features Before Excavation: A Case from the American Southwest." *Journal of Field Archaeology* 17:323–331.

———. 1994a. *Turquoise Ridge and Late Prehistoric Mobility in the Desert Mogollon Region*. Salt Lake City: University of Utah Press.

———. 1994b. "Moving out of the Archaic on the Edge of the Southwest." *American Antiquity* 59:622–638.

Whalen, Michael E. and Paul E. Minnis. 2001. "Architecture and Authority in the Casas Grandes Area, Chihuahua, Mexico." *American Antiquity* 66:651–668.

Wheat, Joe Ben. 1955. *Mogollon Culture Prior to A.D. 1000*. Memoirs of the Society for American Archaeology No. 10. Menasha: Society for American Archaeology.

White, Alan S. 1985. "Presettlement Regeneration Patterns in a Southwestern Ponderosa Pine Stand." *Ecology* 66:589–594.

White, Robert R. 1994. *Hydrology of the Estancia Basin, Central New Mexico*. Water Resources Investigation Report 93–4163. Albuquerque: U.S. Geological Survey.

Whitehead, Harriet. 1993. "Morals, Models, and Motives in a Different Light: A Rumination on Alan P. Fiske's Structure of Social Life." *Ethos* 21:319–356.

Whiteley, Peter M. 1988. *Deliberate Acts: Changing Hopi Culture Through the Oraibi Split*. Tucson: University of Arizona Press.

———2008. *The Orayvi Split: A Hopi Transformation*. American Museum of Natural History Anthropological Papers 87. Washington, DC: U.S. Government Printing Office.

Wiener, Annette. 1985. "Inalienable Wealth." *American Ethnologist* 12:210–227.

———. 1992. *Inalienable Possessions: The Paradox of Keeping While Giving*. Berkeley: University of California Press.

Wiessner, Polly. 2002. "The Vines of Complexity: Egalitarian Structures and the Institutionalization of Inequality Among the Enga." *Current Anthropology* 43: 233–269.

Wilk, Richard and R. M. Netting. 1984. "Households: Changing Forms and Functions." In *Households: Comparative and Historical Studies of the Domestic Group*, edited by R. M. Netting, R. Wilk, and E. Arnold, pp. 1–28. Berkeley: University of California Press.

Wilk, Richard R. and William J. Rathje. 1982. "Household Archaeology." *American Behavioral Scientist* 25:617–639.

Willey, Gordon R. and Philip Phillips. 1958. *Method and Theory in American Archaeology*. Chicago: University of Chicago Press.

Wills, W. H. 1988a. "Agriculture and Sedentism in the American Southwest: Evidence and Interpretations." *Journal of World Prehistory* 2:445–488.

———. 1988b. *Early Prehistoric Farming*. Santa Fe: School of American Research.

———. 1991. "Organizational Strategies and the Emergence of Prehistoric Villages in the American Southwest." In *Between Bands and States*, edited by Susan A. Gregg, pp. 161–180. Occasional Paper No. 9, Center for Archaeological Investigations. Carbondale: Southern Illinois University.

———. 1992. "Foraging Systems and Plant Cultivation During the Emergence of Agricultural Economies in the Prehistoric American Southwest." In *Transitions to Agriculture in Prehistory*, edited by Ann B. Gebauer and T. Douglas Price, pp. 153–176. Monographs in World Archaeology No. 4. Madison: Prehistory Press.

———. 2001. "Pithouse Architecture and the Economics of Household Formation in the Prehistoric American Southwest." *Human Ecology* 29:477–500.

———. 2007. "The Structural Role of Pithouses: Architectural Variation in the Mogollon Highlands, A.D. 100 to 1300." In *Exploring Variability in Mogollon Pithouses*, edited by Barbara J. Roth and Robert J. Stokes, pp. 93–108. Anthropological Research Papers No. 58. Tempe: Arizona State University.

Wills, W. H. and Patricia L. Crown. 2004. "Commensal Politics in the Prehispanic Southwest: An Introductory Review." In *Identity, Feasting, and the Archaeology of the Greater Southwest*. Proceedings of the 2002 Southwest Symposium, edited by Barbara J. Mills, pp. 153–172. Boulder: University Press of Colorado.

Wills, W. H. and Robert D. Leonard, eds. 1994. *The Ancient Southwestern Community*. Albuquerque: University of New Mexico Press.

Wills, W. H. and Thomas C. Windes. 1989. "Evidence for Population Aggregation and Dispersal During the Basketmaker III Period in Chaco Canyon, New Mexico." *American Antiquity* 54:347–369.

Wills, W. H., F. Scott Worman, Wetherbee Dorshow, and Heather Richards-Rissetto. 2012. "Shabik'eschee Village in Chaco Canyon: Beyond the Archetype." *American Antiquity* 77:326–351.

Wilshusen, Richard H. 1991. "Early Villages in the American Southwest: Cross-Cultural and Archaeological Perspectives." Ph.D. dissertation. Department of Anthropology, University of Colorado, Boulder.

Wilshusen, Richard H. and Scott G. Ortman. 1999. "Rethinking the Pueblo I Period in the San Juan Drainage: Aggregation, Migration, and Cultural Diversity." *Kiva* 64:369–399.

Wilshusen, Richard H., Gregson Schachner, and James R. Allison, eds. 2012. *Crucible of Pueblos: The Early Pueblo Period in the Northern Southwest*. Los Angeles: Cotsen Institute of Archaeology Press.

Wilson, John P. 1993. "Quarai: A Turbulent History." In *Salinas: Archaeology, History, and Prehistory*, edited by David Grant Noble, pp. 20–25. Santa Fe: Ancient City Press.

Wilson, John P., Robert H. Leslie, and A. H. Warren. 1983. "Tabira: Outpost on the East." In *Collected Papers in Honor of Charlie R. Steen*, edited by Nancy Fox, pp. 87–158. Papers of the Archaeological Society of New Mexico, No. 8. Albuquerque: The Archaeological Society of New Mexico.

Wiseman, Regge N. 1986. *An Initial Study of the Origins of Chupadero Black-on-White.* Technical Note No. 2. Albuquerque: The Albuquerque Archaeological Society.

Wimmer, Andreas. 2008. "The Making and Unmaking of Ethnic Boundaries: A Multilevel Process Theory." *American Journal of Sociology* 113(4):970–1022.

Woodbury, Richard. 1993. "Early Prehistory of the Estancia Basin." In *Salinas: Archaeology, History, and Prehistory*, edited by David Grant Noble, pp. 2–5. Santa Fe: Ancient City Press.

Works, Martha A. 1992. "Creating Trading Places on the New Mexican Frontier." *Geographical Review* 82:268–281.

Yanagisako, Sylvia J. 1979. "Family and Household: The Analysis of Domestic Groups." *Annual Review of Anthropology* 8:161–205.

Yeager, Jason and Marcello A. Canuto. 2000. "Introducing an Archaeology of Communities." In *The Archaeology of Communities: A New World Perspective*, edited by Marcello A. Canuto and Jason Yaeger, pp. 1–15. New York: Routledge.

Yonts, C. D. and D. S. Nuland. 1984. *Irrigating Dry Beans.* NebGuide G84-686. Lincoln: University of Nebraska.

Zeisel, John. 1984. *Inquiry by Design: Tools for Environment-Behaviour Research.* Cambridge: Cambridge University Press.

Index

About the Author

Alison E. Rautman is an associate professor in the Center for Integrative Studies at Michigan State University and a Registered Professional Archaeologist. Her archaeological research in the Salinas region of central New Mexico has been funded by grants from the National Science Foundation, the Wenner-Gren Foundation for Anthropological Research, The National Geographic Society, Sigma Xi, and the American Council of Learned Societies; she has also received fellowships from the American Association of University Women. In addition to her Salinas research, she has published on issues of historic and ancient cannibalism, ceramic petrography, risk and decision-making in prehistory, gender roles, and anthropological theory. Her publications also include the edited volume *Reading the Body: Representations and Remains in the Archaeological Record* (University of Pennsylvania Press, 2000). From 2009 to 2012, she served as editor of *American Antiquity*, the journal of the Society for American Archaeology.